EPA ASSESSMENT OF RISKS FROM RADON IN HOMES

June 2003

Office of Radiation and Indoor Air
United States Environmental Protection Agency
Washington, DC 20460

PREFACE

Radon has been classified as a known human carcinogen and has been recognized as a significant health problem by groups such as the Centers for Disease Control, the American Lung Association, the American Medical Association, and the American Public Health Association. As such, risks from in-home radon exposure have been a major concern for the EPA. In 1992, EPA published its *Technical Support Document for the 1992 Citizen's Guide to Radon*, which included a description of its methodology for estimating lung cancer risks in the U.S. associated with exposure to radon in homes. That methodology was primarily based on reports published by the National Academy of Sciences (NAS). In one of those reports, known as "BEIR IV" (NAS 1988), a model was derived for estimating the risks from inhaled radon progeny, based on an analysis of epidemiologic results on 4 cohorts of occupationally exposed underground miners. In 1994, the EPA sponsored another study, "BEIR VI", to incorporate additional information that had become available from miner cohort and residential studies. In early 1999, the NAS published its "BEIR VI" report (NAS 1999), which presented new risk models based on information from 11 miner cohorts. A major conclusion of the BEIR VI report was that radon is the second leading cause of lung cancer after smoking.

In light of findings and recommendations in BEIR VI, this report presents a revised risk assessment by EPA's Office of Radiation and Indoor Air (ORIA) for exposure to radon in homes. In response to a request by ORIA, the Radiation Advisory Committee (RAC) of the Science Advisory Board (SAB) has reviewed the methodology used in this report for estimating cancer risks from radon. An initial advisory, finalized in July, 1999 (SAB 1999), found the methodology to be generally acceptable but included recommendations for some adjustments. The RAC met again in November,1999 to consider ORIA's response to their recommendations. The RAC report (SAB 2000) concluded that "ORIA has produced a credible risk assessment and has responded well to the recommendations provided by the RAC in its Advisory." They also offered additional comments and suggestions. Responses to those comments were provided in a letter of October 5, 2000 from Robert Perciasepe, Assistant Administrator of the Office of Air and Radiation.

This report was prepared by EPA staff members David J. Pawel and Jerome S. Puskin, ORIA, Radiation Protection Division. The authors gratefully acknowledge the invaluable assistance provided by Christopher B. Nelson, the constructive review conducted by the RAC, and helpful review comments by Dr. Nancy Chiu and Dr. William Brattin.

The mailing address for the authors is:
      U.S. Environmental Protection Agency
      Office of Radiation and Indoor Air (6608J)
      Washington, DC 20460

ABSTRACT

Background. The U.S. Environmental Protection Agency (EPA) updates its assessment of health risks from indoor radon, which the National Academy of Sciences (NAS) has determined to be the second leading cause of lung cancer after cigarette smoking. This risk assessment is based primarily on results from a recent study of radon health effects (BEIR VI) by the NAS, with some technical adjustments and extensions. In BEIR VI, the NAS projected 15,400 or 21,800 excess lung cancer deaths in the U.S. each year, using two preferred risk models developed from data from 11 cohorts of miners.

Methods. EPA modified and extended the approach used in BEIR VI. First, a single model is constructed that yields numerical results midway between what would be obtained using the two BEIR VI preferred models. Second, noting that the BEIR VI definition of excess risk effectively omits premature deaths caused by radon in people who would otherwise have eventually died of lung cancer, EPA modifies the BEIR VI calculations to include all radon-induced lung cancer deaths. Third, EPA uses more detailed smoking prevalence data and more recent mortality data for its calculations than was used in BEIR VI. Fourth, whereas BEIR VI estimated the fractional increase in lung cancers due to radon, EPA also provides numerical estimates of the risk per unit exposure [lung cancer deaths per working level month (WLM)].

Results. Based on its analysis, EPA estimates that out of a total of 157,400 lung cancer deaths nationally in 1995, 21,100 (13.4%) were radon related. Among NS, an estimated 26% were radon related. Estimates of risk per unit exposure are 5.38×10^{-4} per WLM for the U.S. population; 9.68×10^{-4}/WLM for ever smokers (ES); and 1.67×10^{-4} per WLM for never smokers (NS). The estimated risks from lifetime exposure at the 4 pCi/L action level are: 2.3% for the entire population, 4.1% for ES, and 0.73% for NS. A Monte Carlo uncertainty analysis that accounts for only those factors that can be quantified without relying too heavily on expert opinion indicates that estimates for the U.S. population and ES may be accurate to within factors of about 2 or 3.

Conclusions. The effects of radon and cigarette smoking are *synergistic*, so that smokers are at higher risk from radon. Consequently, if projected reductions in U.S. smoking rates hold up, some decrease in radon-induced lung cancers is expected, concomitant with decreases in lung cancer, generally; nevertheless, it is anticipated that indoor radon will remain an important public health problem, contributing to thousands of lung cancer deaths annually.

CONTENTS

Section	Page

LIST OF TABLES

LIST OF FIGURES

EXECUTIVE SUMMARY

Radon-222 is a noble gas produced by radioactive decay of radium-226, which is widely distributed in soils and rocks. Radon-222 decays into a series of short-lived radioisotopes. These decay products are often referred to as *radon progeny* or *daughters*. Because it is chemically inert, most inhaled radon is rapidly exhaled, but the inhaled decay products readily deposit in the lung, where they irradiate sensitive cells in the airways, thereby enhancing the risk of lung cancer.

In 1999, the National Research Council of the National Academy of Sciences published the BEIR VI report, *Health Effects of Exposure to Radon* (NAS 1999), which assessed the risks to the U.S. population from radon in homes. The authors of this study, sponsored by the EPA, had the benefit of extensive new information not available to the authors of the Academy's previous BEIR IV report on the risks from radon and other alpha emitters (NAS 1988). On the basis of epidemiologic evidence from miners and an understanding of the biologic effects of alpha radiation, the committee concluded that residential exposure to radon is expected to be a cause of lung cancer in the population. Based on a statistical analysis of epidemiologic data on 11 cohorts of occupationally exposed underground miners, the committee developed two preferred risk models from which they projected, respectively, 15,400 or 21,800 excess lung cancer cases in the U.S. each year. An analysis of the uncertainties suggested a range of 3,000 to 33,000 cases per year. The committee concluded that "this indicates a public health problem and makes indoor radon the second leading cause of lung cancer after cigarette smoking."

Both of the preferred BEIR VI models are framed in terms of excess relative risk (ERR), which represents the fractional increase in lung cancer risk due to a specified exposure.[1] To estimate the risk at any given age from a past exposure, one multiplies the ERR times the baseline lung cancer rate for an individual of that age (and, if appropriate, sex or smoking category). The lifetime risk from an arbitrary exposure can be calculated using a specified risk model in conjunction with life-table methods that incorporate competing causes of death. In both of these BEIR VI models the ERR falls off with time-since-exposure and with age at risk; nevertheless, because of the increasing baseline rate of lung cancer with age, the calculated risk from a given exposure often increases with increasing age.

An important finding in BEIR VI, based on updated and expanded miner data, is that risk from a given exposure tends to increase when that exposure is more spread

---

[1] Exposures are measured in units of working level months (WLM), a measure of potential alpha particle energy that will be released by short-lived radon decay products per liter of air.

1

out over time. For the relatively low exposure rates or long time durations of most concern for EPA, the risk per unit (WLM) exposure is maximal and increases linearly with radon exposure.

Another new finding is that the estimated ERR is about twice as high for never smokers (NS) as for ever smokers (ES). Estimates indicate that radon exposure accounts for about 1 in 8 ES lung cancer deaths and 1 in 4 NS lung cancer deaths. However, since ES have a much higher baseline lung cancer rate than NS, the risk of a radon-induced lung cancer, on an absolute scale, is still much higher than for NS.

Although there is a growing body of data from epidemiological (case-control) studies showing a correlation between lung cancer and radon exposures in homes, these results do not conclusively demonstrate an excess risk in homes with elevated radon and are inadequate as a basis for quantitative risk estimation. Thus, estimates of risk for indoor exposures must still be extrapolated using models derived from the miner data. There are a number of important differences between mine and indoor exposures that must be considered in making this extrapolation.

First, due to physical and physiological factors, the alpha particle dose to target cells in the lung per WLM could be higher or lower in the case of residential exposures than for mine exposures. Since the risk is presumed to be proportional to dose, a model derived from the miner data might need to be adjusted to account for these differences. The BEIR VI risk estimates were based on the premise that the effects of these differences approximately counterbalanced each other in such a way that no adjustment was warranted.   Doubts about this premise were expressed by Cavallo (2000). Cavallo correctly noted inconsistencies in portions of BEIR VI relating to how doses from exposures in mines and homes compare, and suggested that as a result the BEIR VI report may have overstated risks from residential exposures. More recently, James et al. (2003) submitted a report which carefully reexamined issues raised by Cavallo. James et al. reaffirmed that the effects on doses of differences between homes and mines do approximately counterbalance each other so that no adjustment would be needed for in-home risk calculations. It follows that the inconsistencies in BEIR VI noted by Cavallo did not lead to an overestimate of the risks from radon.

Second, other agents in the atmosphere of underground mines, such as arsenic, silica, and diesel fumes, could modify the lung cancer risk associated with exposure to radon progeny. BEIR VI cited evidence that the latter two types of exposures were probably not strong modifiers of risk but that arsenic might be a source of positive bias in the risk estimates.

Third, the exposure rates in homes are generally lower than the lowest levels for which we have clear evidence of excess risk in mines. Consequently, assessment of indoor radon risks requires an extrapolation to lower exposure rates. Although the miner data and radiobiological data are both suggestive of a constant risk per unit exposure as one extrapolates downward from the lowest miner exposures, this assumption has been questioned. An ecological study has indicated that lung cancer rates are negatively correlated with average radon concentrations across U.S. counties (Cohen 1995), suggesting that the risks from very low levels of radon have been overestimated, or that such exposure levels might even protective against lung cancer. Biologically based models have also been proposed that could project substantially reduced carcinogenicity at low doses (for example, Moolgavkar and Luebeck 1990, Elkind 1994). Numerous critics, including the BEIR VI committee, have discounted the ecological study results because of methodological limitations, and the biologically based models remain highly speculative. The BEIR VI committee adopted the linear no-threshold assumption based on our current understanding of the mechanisms of radon-induced lung cancer, but recognized that this understanding is incomplete and that therefore the evidence for this assumption is not conclusive.

In this document EPA updates its assessment of the health risks from indoor radon, based primarily on the BEIR VI report, with some technical adjustments and extensions. First, EPA constructs a single model that yields numerical results midway between what would be obtained using the two BEIR VI preferred models. Second, noting that the BEIR VI definition of excess risk effectively omits premature deaths caused by radon in people who would otherwise have eventually died of lung cancer, EPA modifies the BEIR VI calculations so as to include all radon-induced lung cancer deaths. Third, whereas the BEIR VI committee assumed that a fixed percentage of adult males or females were ES, EPA uses age-specific smoking prevalence data. Fourth, whereas BEIR VI estimated the fractional increase in lung cancers due to radon, EPA also provides numerical estimates of the risk per unit exposure [lung cancer deaths per working level month (WLM)] and the number of years of life lost per cancer death.

Based on its analysis, EPA estimates that out of a total of 157,400 lung cancer deaths nationally in 1995, 21,100 (13.4%) were radon related. Although it is not feasible to totally eliminate radon from the air, it is estimated that about one-fourth of the radon-related lung cancers could be averted by reducing radon concentrations in homes that exceed EPA's recommended 4 picocurie per liter (pCi/L) action level (NAS 1999).

It is estimated that 86% of the radon-related lung cancer deaths were in ES, compared to 93% for all lung cancer deaths. The projected average years of life lost are higher for the radon-related cases (17 y) than for lung cancer deaths generally (12 y). Estimates of risk per unit exposure are as follows: 5.38×10^{-4}/WLM (all); 9.68×10^{-4}/WLM (ES); and 1.67×10^{-4}/WLM (NS). Based on an assumed average equilibrium fraction of 40% between radon and its decay products and an indoor occupancy of 70%, the estimated risks from lifetime exposure at the 4 pCi/L action level are: 2.3% (all), 4.1% (ES), and 0.73% (NS). Although estimated absolute risks are much higher for ES than NS, estimated relative risks are higher for NS. It is estimated that among NS about one-quarter (26%) of lung cancers are due to radon compared to about one-eighth (12%) among ES. It was more difficult to estimate risks for current smokers. Because of limitations of the data from the miner cohorts, the BEIR VI models did not specify excess relative risks for current smokers. Estimates of risk for current smokers (calculated by presuming that they start smoking at age 18 y and do not quit) are 1.5×10^{-3} per WLM, or over 6% for a lifetime exposure at 4 pCi/L.

EPA also reexamines the issue of uncertainty in the risk estimates. Emphasizing the uncertainty in extrapolating risk estimates from observations on miners exposed to higher levels of radon than are ordinarily found in homes, BEIR VI derived its preferred uncertainty bounds (95% confidence limits 3,300 to 32,600) using a constant relative risk model obtained by a statistical fit to a restricted set of data on miners exposed to less than 50 WLM — levels that are comparable to lifetime residential exposures. The sampling errors are large with this limited data base; as a consequence the resulting confidence range may be overly broad. EPA adopts an alternative approach, deriving its estimates of uncertainty using the BEIR VI preferred models, with some explicit consideration of model uncertainties. However, like BEIR VI, EPA was unable to quantify all the relevant sources of uncertainty. These uncertainties are discussed qualitatively (or semi-quantitatively) and, for perspective, results of sensitivity analyses for some of these variables are included. From a Monte Carlo analysis of those uncertainties that could be quantified, EPA estimates a 90% subjective confidence interval of 2 to 12 $\times 10^{-4}$ lung cancer deaths per WLM, for the general population. The corresponding 90% interval for radon-induced lung cancer cases in 1995 is 8,000 to 45,000. Since the interval would be wider if additional sources of uncertainty had been accounted for in the analysis, it is plausible that the number of radon-induced deaths is smaller than 8,000 (but unlikely that it would be as small as 3,300). However, given the predominant role smoking is known to play in the causation of lung cancer, it is unlikely that radon accounts for as many as 45,000 deaths or 12 $\times 10^{-4}$ lung cancer deaths per WLM. Risk estimates for exposures to specific subgroups, especially children, NS and former smokers, have a higher degree of uncertainty than estimates for the general population.

The effects of radon and cigarette smoking are *synergistic*, so that smokers are at higher risk from radon. Consequently, if projected reductions in U.S. smoking rates hold up, some decrease in radon-induced lung cancers is expected, concomitant with decreases in lung cancer, generally; nevertheless, it is anticipated that indoor radon will remain an important public health problem, contributing to thousands of lung cancer deaths annually.

I. Introduction

In 1992, EPA published its *Technical Support Document for the 1992 Citizen's Guide to Radon,* which included a description of its methodology for estimating lung cancer risks in the U.S. associated with exposure to radon in homes. That methodology was primarily based on two reports published by the National Academy of Sciences (NAS), referred to here as "BEIR IV" (NAS 1988) and the "Comparative Dosimetry Report" (NAS 1991). In BEIR IV, a model was derived for estimating the risks from inhaled radon progeny, based on an analysis of epidemiologic results on 4 cohorts of occupationally exposed underground miners. In the Comparative Dosimetry Report, estimates of radiation dose to potential target cells in the lung were calculated under mine and residential conditions, respectively. Results were expressed in terms of a ratio, K, representing the quotient of the dose of alpha energy per unit exposure to an individual in a home compared to that for a miner in a mine. It was concluded that the dose per unit exposure was typically about 30% lower in homes than in mines ($K \approx 0.7$), implying a 30% reduction in the risk coefficient applicable to home environments from what would be estimated from miner data.

Subsequently, EPA sponsored another NAS study (BEIR VI), which provided new risk models and estimates of the K-factor, based on much more complete information (NAS 1999). Data on 11 miner cohorts were now available, including further follow-up of the 4 cohorts upon which the BEIR IV model was based. In addition, some new information had become available regarding exposure conditions in mines and homes that led to a revised estimate of K. In response to questions raised about issues relating to the K-factor in BEIR VI (Cavallo 2000), the EPA sponsored a study in which it was concluded that, under the exposure assumptions employed in BEIR VI, the value used for the K- factor was appropriate (James et al. 2003).

EPA is now revising its assessment of risks from indoor radon in light of the findings and recommendations in BEIR VI. The revised methodology includes some extensions and modifications from the approach in BEIR VI. These extensions and modifications were made after an advisory review from the Agency's Radiation Advisory Committee (RAC). Taken together, these adjustments have only a minor impact on the estimated number of radon induced lung cancers occurring each year.

This document will serve as a technical basis for EPA's estimates of risk from radon in homes. It provides estimates of the risk per unit exposure and projects the number of fatal lung cancers occurring in the U.S. population each year due to radon. It also provides separate estimates for males and females, and for ever- and never-smokers. Finally, it discusses the uncertainties in these estimates. It is anticipated that the methodology and results presented here will be used in developing guidance for the members of the public in addressing elevated radon levels in their homes. These results may also be used for regulatory purposes: *e.g.*, to set cleanup levels for radium in soil or to set maximum concentration levels for radon in drinking water.

II. Scientific Background

Radon-222 is a noble gas produced by the radioactive decay of radium-226, which is widely distributed in uranium-containing soils and rocks. The radon readily escapes from the soil or rock where it is generated and enters surrounding water or air. The most important pathway for human exposure is through the permeation of underlying soil gas into buildings, although indoor radon can also come from water, outside air, or building materials containing radium. Radon-222 decays with a half-life of 3.82 days into a series of short-lived radioisotopes collectively referred to as *radon daughters* or *progeny*. Since it is chemically inert, most inhaled radon-222 is rapidly exhaled, whereas inhaled progeny readily deposit in the airways of the lung. Two of these daughters, polonium-218 and polonium-214, emit alpha-particles. When this happens in the lung, the radiation can damage the cells lining the airways, leading ultimately to cancer. (Nuclear decay of radon decay products also releases energy in the form of beta particles and high energy photons, but the biological damage resulting from these emissions is believed to be small compared to that from alpha particles.)

Two other radon isotopes – radon-219 (actinon), and radon-220 (thoron) – occur in nature and produce radioactive radon daughters. Because of its very short half-life (3.9 s), environmental concentrations of actinon and its daughters are extremely low, so their contribution to human exposure is negligible. The half-life of thoron is also relatively short (56 s), and a lower fraction of released alpha-particle energy is absorbed within target cells in the bronchial epithelium than in the case of radon-222. As a result, thoron is thought to pose less of a problem than radon-222, but we have rather limited information on human exposure to thoron, and no direct information on its carcinogenicity in humans . For the remainder of this document, we shall focus only on radon-222 and its daughters. Following common usage, the term *radon* will in some cases refer simply to radon-222, but sometimes to radon-222 *plus* its progeny. For example, one often talks about "radon risk" when most of that risk is actually conferred by inhaled decay products.

Radon concentrations in air are commonly expressed in picocuries per liter (pCi/L) in the U.S., but in western Europe, they are given in SI units of bequerels per cubic meter (Bq/m^3), where a Bq is 1 nuclear disintegration per second. By definition, 1 picocurie is equal to 0.037 Bq; hence, 1 pCi/L corresponds to 37 Bq/m^3.

Radon progeny concentrations are commonly expressed in working levels (WL). One WL is defined as any combination of short-lived radon daughters in 1 liter of air that results in the ultimate release of 1.3×10^5 million electron volts of alpha energy. If a closed volume is constantly supplied with radon, the concentration of short-lived daughters will increase until an equilibrium is reached where the rate of decay of each daughter will equal that of the radon itself. Under these conditions each pCi/L of radon will give rise to (almost precisely) 0.01 WL. Ordinarily these conditions do not hold: in homes, the *equilibrium fraction* is typically 40%; *i.e.*, there will be 0.004 WL of progeny for each pCi/L of radon in air (NAS 1999).

Cumulative radon daughter exposures are measured in working level months (WLM), a unit devised originally for occupational applications. Exposure is proportional to concentration (WL) and time, with exposure to 1 WL for 170 h being defined as 1 WLM. To convert from residential exposures expressed in pCi/L, the BEIR VI committee assumed that the fraction of time spent indoors is 70%. It follows that an indoor radon concentration of 1 pCi/L would on average result in an exposure of 0.144 WLM/y = (1 pCi/L) [(0.7)(0.004) WL/(pCi/L)] (51.6 WLM/WL-y).

There is overwhelming evidence that exposure to radon and its decay products can lead to lung cancer. Since the 1500s, it has been recognized that underground miners in the Erz mountains of eastern Europe are susceptible to high mortality from respiratory disease. In the late 1800s and early 1900s, it was shown that these deaths were due to lung cancer. The finding of high levels of radon in these mines led to the hypothesis that it was responsible for inducing cancer. This conclusion has been confirmed by numerous studies of radon-exposed underground miners and laboratory animals.

The most important information concerning the health risks from radon comes from epidemiological studies of underground miners. In these "cohort" studies, lung cancer mortality is monitored over time in a group of miners and correlated with the miners' estimated past radon exposure. The BEIR VI committee analyzed results from 11 separate miner cohorts, each of which shows a statistically significant elevation in lung cancer mortality with increasing radon exposure. Summary information on the epidemiologic follow-up of the 11 cohorts is provided in Table 1.

Table 2 summarizes information on the miners' exposure and the excess relative risk (ERR) per unit exposure in each cohort. The ERR represents the multiplicative increment to the excess lung cancer mortality beyond background resulting from the exposure. From Table 2 it is clear that there is heterogeneity in the estimates of the ERR per unit exposure derived from the various studies. Some of this heterogeneity is attributable to random error, and some to exposure rate or age and temporal parameters discussed below. There is, however, unexplained residual heterogeneity, possibly due to systematic errors in exposure ascertainment, unaccounted for differences in the study populations (genetic, lifestyle, etc.), or confounding mine exposures.

Table 1: Miner cohorts, number exposed, person-years of epidemiologic follow-up,and lung cancer deaths (NAS 1999).

Study	Type of Mine	Number of Workers	Number of person-years	Number of lung cancers
China	Tin	13,649	134,842	936
Czechoslovakia	Uranium	4,320	102,650	701
Colorado Plateau[a]	Uranium	3,347	79,536	334
Ontario	Uranium	21,346	300,608	285
Newfoundland	Fluorspar	1,751	33,795	112
Sweden	Iron	1,294	32,452	79
New Mexico	Uranium	3,457	46,800	68
Beaverlodge (Canada)	Uranium	6,895	67,080	56
Port Radium (Canada)	Uranium	1,420	31,454	39
Radium Hill (Australia)	Uranium	1,457	24,138	31
France	Uranium	1,769	39,172	45
Total[b]		60,606	888,906	2,674

[a] Exposure limited to <3,200 WLM.
[b] Totals adjusted for miners and lung cancers included under both Colorado and New Mexico studies.

Table 2: Miner cohorts, radon exposure, and estimates of excess relative risk per WLM exposure with 95% CI (NAS 1999).

Study	Mean WLM[a]	Mean duration (y)	Mean WL[a]	ERR/WLM %
China	286.0	12.9	1.7	0.16 (0.1-0.2)
Czechoslovakia	196.8	6.7	2.8	0.34 (0.2-0.6)
Colorado Plateau	578.6	3.9	11.7	0.42 (0.3-0.7)
Ontario	31.0	3.0	0.9	0.89 (0.5-1.5)
Newfoundland	388.4	4.8	4.9	0.76 (0.4-1.3)
Sweden	80.6	18.2	0.4	0.95 (0.1-4.1)
New Mexico	110.9	5.6	1.6	1.72 (0.6-6.7)
Beaverlodge	21.2	1.7	1.3	2.21 (0.9-5.6)
Port Radium	243.0	1.2	14.9	0.19 (0.1-1.6)
Radium Hill	7.6	1.1	0.7	5.06 (1.0-12.2)
France	59.4	7.2	0.8	0.36 (0.0-1.2)
Total	164.4	5.7	2.9	

[a] Weighted by person-years; includes 5-year lag period.

III. Previous Methodology for Calculating Risks

EPA's previous methodology for calculating the risks from indoor radon exposures was described in the *Technical Support Document for the 1992 Citizen's Guide to Radon* (EPA 1992). That methodology made use of the risk model derived in the 1988 National Academy of Sciences' BEIR IV Report, based on a statistical analysis of results from four epidemiologic studies of radon-exposed underground miners (NAS 1990). The preferred model in the BEIR IV Report expresses the excess relative risk (ERR) of lung cancer death at age *a,* as a function of past exposure:

$$ERR(a) = 0.025 \, \gamma(a) \, (W_1 + \tfrac{1}{2} W_2) \tag{1}$$

where $\gamma(a)$ is an age-specific adjustment to the relative risk coefficient, as follows:

$$\gamma(a) = 1.2 \text{ when } a < 55 \text{ y}$$
$$= 1.0 \text{ when } 55 \text{ y} \le a < 65 \text{ y}$$
$$= 0.4 \text{ when } a \ge 65 \text{ y}$$

W_1 is the cumulative exposure received 5-15 y before age a, and W_2 is the cumulative exposure up to age a-15. Thus, the model incorporates a fall-off in the ERR with age at expression and, independently, with time-since-exposure.

In extrapolating risk estimates from mine to home exposures, EPA, NAS and others have assumed that the risk is proportional to the dose to target cells lining the airways of the lung. Thus, in order to estimate risk from home exposures, the right-hand side of Equation 1 is multiplied by a factor K, which is equal to the ratio of the dose per WLM exposure in homes relative to mines. Numerous parameters affect estimates of the dose per WLM and, therefore, K. These include breathing rates, location of target cells in the lung, mucus thickness and mucocilliary clearance rates, the size distribution of aerosol particles to which radon decay products are attached, the relative concentrations of radon decay products, and the proportion of decay products existing as an unattached (ultrafine) fraction. The BEIR IV committee concluded that K was reasonably close to 1 and recommended that Equation 1 be applied for the case of residential exposures. A subsequent NAS committee examined this issue in greater depth and determined that a best estimate for K was about 0.7 (NAS 1991). Accordingly, EPA adopted the following risk model for residential exposures (EPA 1992):

$$ERR(a) = 0.0175 \, \gamma(a) \, (W_1 + \tfrac{1}{2} W_2) \tag{2}$$

The risk of a radon-induced lung cancer death at age a was then calculated as the product of $ERR(a)$ times the baseline lung cancer mortality rate at age a. With the aid of life-table techniques (EPA 1992), the average risk to a member of the 1989-91 life-table population was found to be approximately 2.24×10^{-4} per WLM. Using this value in conjunction with an estimated annual average exposure in the U.S. of 0.242 WLM/y, the number of radon-induced lung cancer deaths each year in a population of 250 million was estimated to be 13,600. In that report, EPA employed a correction that subtracted off the estimated radon-induced lung cancer deaths occurring at each age from the reported lung cancer mortality. This "baseline correction" had the effect of reducing the population risk estimate by about 10%.

Consistent with the limited evidence available at the time of the BEIR IV Report's publication, the model assumed a multiplicative interaction between smoking and radon exposure; consequently, the ERR was independent of smoking status. Also, while there was some indication of an increased risk at low exposure rates and longer exposure durations in the Colorado Plateau miners, these effects were not consistent across the four cohorts analyzed. As a result, the BEIR IV committee assumed that the risk was not explicitly dependent on exposure rate or duration.

Soon after publication of BEIR IV, the International Council for Radiological Protection (ICRP) published ICRP Report 65 (ICRP 1993), which relied on essentially the same data as in BEIR IV. ICRP's risk projection model was also a relative risk model that depended both on time-since-exposure and age at exposure, but not exposure rate or duration.

IV. BEIR VI Risk Models

A. Statistical Fits to the Miner Data

In 1998, the NAS published a new report, BEIR VI, that updated the findings on radon risk presented in BEIR IV. Two preferred models were developed by the BEIR VI committee based on a combined statistical analysis of results from the latest epidemiologic follow-up of 11 cohorts of underground miners, which, in all, included about 2,700 lung cancers among 68,000 miners, representing nearly 1.2 million person-years of observations. Both preferred BEIR VI models, like the preferred model in BEIR IV, incorporate a 5-y minimum latency period and a fall-off in the ERR with age at expression and time-since-exposure, but the BEIR VI models provide a more detailed break-down of the risk for ages over 65 y and times since exposure greater than 15 y.

Unlike what was found with the more limited BEIR IV and ICRP analyses, the BEIR VI committee was able to conclude that the ERR per WLM increased with decreasing exposure rate or with increasing exposure duration (holding cumulative exposure constant). To account for this "inverse dose rate" effect, the committee introduced a parameter dependent on the radon-daughter working level (WL) concentration or, alternatively, the duration of exposure. Respectively, this gave rise to the two alternative preferred models – the "exposure-age-concentration model" and the "exposure-age-duration model." For brevity, these will generally be referred to here as the "concentration" and "duration" models.

Mathematically, the ERR in the two models can be represented as:

$$ERR = \beta \ (w_{5-14} + \theta_{15-24} \ w_{15-24} + \theta_{25+} \ w_{25+}) \phi_{age} \gamma_z \qquad (3)$$

where. β is the exposure-response parameter (risk coefficient); the exposure windows, w_{5-14}, w_{15-24} and w_{25+}, define the exposures incurred 5-14 y, 15-24 y and ≥ 25 y before the current age; and θ_{15-24} and θ_{25+} represent the relative contributions to risk from exposures 15-24 y and ≥ 25 y before the attained age. The parameters ϕ_{age} and γ_z define effect-modification factors representing, respectively, multiple categories of attained age (ϕ_{age}) and of either exposure rate or exposure duration (γ_z). The values for these parameters are summarized in Table 3.

Table 3: Parameter estimates for BEIR VI models (NAS 1999).

Duration Model		Concentration Model	
$\beta \times 100$	0.55	$\beta \times 100$	7.68
Time-since-exposure			
θ_{15-24}	0.72	θ_{15-24}	0.78
θ_{25+}	0.44	θ_{25+}	0.51
Attained age			
$\phi_{<55}$	1.00	$\phi_{<55}$	1.00
ϕ_{55-64}	0.52	ϕ_{55-64}	0.57
ϕ_{65-74}	0.28	ϕ_{65-74}	0.29
ϕ_{75+}	0.13	ϕ_{75+}	0.09
Duration of exposure		Exposure rate (WL)	
$\gamma_{<5}$	1.00	$\gamma_{<0.5}$	1.00
γ_{5-14}	2.78	$\gamma_{0.5-1}$	0.49
γ_{15-24}	4.42	γ_{1-3}	0.37
γ_{25-34}	6.62	γ_{3-5}	0.32
γ_{35+}	10.2	γ_{5-15}	0.17
		γ_{15+}	0.11

B. Extrapolation from Mines to Homes

The analysis of the miner studies provides models for estimating the risk per unit exposure, as a function of age-at-expression, time-since-exposure, and exposure rate or duration. However, exposure conditions in homes differ from those in mines, with respect to both the physical properties of the inhaled radon decay products and the breathing patterns in the two environments. Using the terminology employed in the NAS "BEIR IV" and "Comparative Dosimetry" reports (NAS 1988, 1991), the risk per unit exposure in homes, $(Risk)_h / (WLM)_h$, can be related to that in mines, $(Risk)_m / (WLM)_m$, by a dimensionless factor, K,

$$K = \frac{(\text{Risk})_h / (\text{WLM})_h}{(\text{Risk})_m / (\text{WLM})_m}$$

In extrapolating from mine to residential conditions, it is assumed that the risk is proportional to the alpha particle dose delivered to sensitive target cells in the bronchial epithelium. Then, K can be written as the ratio of the estimated doses per unit exposure in the two environments:

$$K = \frac{(\text{Dose})_h / (\text{WLM})_h}{(\text{Dose})_m / (\text{WLM})_m}$$

Previously, the NAS estimated that the dose from residential exposures was typically 30% lower than from an equal WLM exposure in mines (NAS 1991); hence, EPA applied a K-factor of 0.7 in calculating the risk in homes based on the models derived from miner studies (EPA 1992).

In BEIR VI the NAS derived a revised estimate of K equal to 1. The most important changes in assumptions from the previous report was a reduction in the breathing rate for miners and an increase in the size of particles associated with mine exposures. However, in BEIR VI, the K-factor was defined in terms of *radon gas* rather than *radon daughter* exposure (NAS 1999, Appendix B). This value appeared to have been misapplied in projecting risk from radon exposure in homes (Cavallo 2000). Under the sponsorship of EPA, James has reexamined the issue and concluded that, under the exposure assumptions employed in BEIR VI, a "best estimate" of K – as properly defined by the equation above – is in fact approximately 1 (James et al. 2003). Hence, the risk projections made for residential exposures in BEIR VI do not require modification (James et al. 2003, Krewski et al. 2002). Nominal estimates of risk for residential exposures in this report are therefore also calculated using a value of K equal to 1.

C. Smoking

The BEIR VI committee had smoking information on five of the miner cohorts, from which it concluded that there was a submultiplicative interaction between radon and smoking in causing lung cancer. That is, the ERR per WLM was higher for never smokers[2] (NS) than for ever smokers (ES), although the absolute risk per WLM was still much higher in the latter, given their much higher rate of lung cancer. The data on never-smoking miners are rather limited, and there is considerable uncertainty in the

---

[2]Never smokers are defined as those persons who had not yet smoked 100 cigarettes; ever smokers include all those who had smoked 100 cigarettes or more.

magnitude of the risk among this group. As a best estimate, the BEIR VI committee determined that the NS should be assigned a relative risk coefficient (β) twice that for the general population, in each of the two models defined above. For consistency, the value of β for ES in the respective models was adjusted downward by a factor of 0.9 from that for the general population.

D. Calculation of Attributable Risk and Lung Cancer Deaths

The two NAS preferred models described above can be used to estimate lung cancer risks in any population for which radon exposure rates and vital statistics can be specified. As will be seen in a Section VI.C., the fraction of lung cancer deaths due to radon — referred to in BEIR VI as the attributable risk (AR) — is only weakly dependent on lung cancer rates in the population. The BEIR VI committee chose to focus primarily on AR calculations. Unlike BEIR IV, the BEIR VI report contains no estimate of the lifetime risk per WLM, which would be a strong function of the lung cancer rate in the population.

The BEIR VI committee first calculated AR for sub-populations of male and female ES and NS. For this calculation, they presumed a steady state population governed by 1985-1989 mortality rates and an average annual exposure of 0.181 WLM/y. The exposure estimate was based on: (1) an average residential radon level of 1.25 pCi/L derived from EPA's National Residential Radon Survey (Marcinowski *et al.* 1994); (2) an estimated average equilibrium fraction (*F*) of 40%; and (3) an assumed 70% occupancy factor (Ω), representing the estimated fraction of time spent indoors at home by the population. The age-specific mortality rates for ES and NS were modified from those for the general population to account for the higher lung cancer mortality in ES. For males, the age-specific lung cancer rate for ES was taken to be 14 times that for NS; for females, the ratio was assumed estimated to be 12. It was further estimated that, among adults, 58% of all males and 42% of all females are ES (independent of age).

The attributable risks estimated in this way by the BEIR VI committee are given in Table 4.

Table 4: Estimated AR for domestic radon exposure using 1985-1989 U.S. population mortality rates (NAS 1999).

Model	Population	ES	NS
Males			
Concentration	0.141	0.125	0.258
Duration	0.099	0.087	0.189
Females			
Concentration	0.153	0.137	0.269
Duration	0.108	0.096	0.197

Assuming that 95% and 90% of all lung cancers in males and females, respectively, occur in ES and that the attributable risks are applicable to the 1995 U.S. population, radon-attributable lung cancer deaths were estimated for that year by the NAS. The results are given in Table 5.

Table 5: Estimated number of lung cancer deaths in the U.S. in 1995 attributable to indoor residential radon progeny exposure (NAS 1999).

Smoking Status	Lung Cancer Deaths	Radon-Attributable Lung Cancer Deaths	
		Concentration Model	Duration Model
Males			
Total	95,400	12,500	8,800
Ever smokers	90,600	11,300	7,900
Never smokers	4,800	1,200	900
Females			
Total	62,000	9,300	6,600
Ever smokers	55,800	8,300	5,400
Never smokers	6,200	1,700	1,200
Males and Females			
Total	157,400	21,800	15,400
Ever smokers	146,400	18,900	13,300
Never smokers	11,000	2,900	2,100

V. Residential Studies

Two types of epidemiologic studies of the association between lung cancer and radon exposure in homes have been performed and are reviewed in BEIR VI: ecologic and case-control. In the former, variations in average radon levels between geographic areas are correlated with corresponding variations in lung cancer rates. In the latter, measured radon levels in the homes of lung cancer cases are compared with those of control subjects who do not have the disease.

The most extensive ecologic study has been carried out by Cohen, who collected a large data base of short-term radon measurements in residences across the U.S. (Cohen 1990, 1995). Grouping the data by county, Cohen found a *negative* correlation between average radon level and age-adjusted lung cancer rate. This has led some to conclude that radon, at typical indoor levels, presents no risk for lung cancer.

A number of criticisms have arisen regarding this use of an ecologic study (NAS 1999). Aside from the biological implausibility of the results and the apparent disagreement with the results from miner cohort studies and residential case-control studies (see below), the most serious of these revolve around the question of possible confounding with smoking, which contributes to a very high percentage of lung cancer cases. In particular, if radon levels were inversely correlated with smoking across counties, it is easy to see that one can have a spurious inverse correlation between average radon level and lung cancer rate. A more subtle bias can arise from the synergism between radon and smoking in causing lung cancer if smoking and radon levels are correlated *within counties* (Greenland and Robins 1994, Lubin 1998). Cohen has argued that the likely magnitude of these kinds of biases is too small to explain his negative correlation, and the controversy continues (Smith *et al.* 1998, Cohen 1998, Cohen 1998a, Lubin 1998a, Field *et al.* 1998, Goldsmith 1999). The BEIR VI committee sided with the critics and concluded that Cohen's inverse correlation "was considered to have resulted from inherent limitations of the ecologic method" and "was considered to be an inappropriate basis for concluding that indoor radon is not a potential cause of lung cancer." Most recently, Puskin (2003) found that Cohen's radon levels have quantitatively similar, strongly negative correlations with cancer rates for cancers strongly linked to cigarette smoking, weaker negative correlations for certain cancers weakly dependent on smoking, and no such correlation for cancers not linked to smoking. These results support the hypothesis that the negative trend reported by Cohen for lung cancer can be largely accounted for by a negative correlation between smoking and radon levels across counties.

Numerous case-control studies of radon exposure and lung cancer were begun in recent years, and most are now either completed or nearing completion. A meta-analysis of eight published case-control studies showed an enhanced risk for lung cancer associated with elevated radon exposure, but the enhancement was barely statistically significant (Lubin and Boice 1996, NAS 1999). The lack of significance is not surprising in view of the limited statistical power achievable at the modestly elevated radon levels generally found in homes. Indeed the observed excess risk is very close to what is expected based on the miner data; moreover, the results deviate significantly from a projection based on the ecologic data discussed above (NAS 1999). Additional results from case-control studies have been reported subsequent to the BEIR VI analysis that provide further support for an increase in lung cancer risk due to radon exposure in homes (Lubin 1999).

VI. Methodology for Calculating Radon Risk

A. Overview

Described here is the newly developed EPA method for calculating lifetime radon-related risk estimates based on the findings of BEIR VI. These include estimates of the *etiologic* fraction (radon-induced fraction of lung cancer deaths), the lifetime risk per

WLM (probablility of a radon-induced cancer death), years of life lost (*YLL*) per (radon-induced) cancer death, and numbers of radon-induced cancer deaths per year. The BEIR VI committee provided estimates of numbers of excess lung cancer deaths and the excess fraction of lung cancer deaths due to radon exposure, but did not provide estimates of risk per WLM or *YLL* per cancer death. Their estimates were based on two different models for relative risk: "the concentration model" and "the duration model," as described in Section IV.A. The concentration model risk estimates were about 40% higher than the duration model estimates. As discussed below in Section B, EPA is basing its estimates on a scaled version of the BEIR VI concentration model. The scaling results in estimated numbers of lung cancer deaths intermediate between the BEIR VI concentration and duration model estimates. Other refinements and extensions to the BEIR VI analysis to meet EPA's needs include:

1) The BEIR VI committee used life-table methods to calculate their risk estimates for NS and ES. These estimates were based on the assumption that 58% of adult males and 42% of adult females are ES, regardless of age. EPA uses age-specific smoking prevalence data (DHHS 1997) shown in Appendix A.

2) The BEIR VI committee calculated the "excess risk" or the increase in the probability of dying from a lung cancer. EPA uses an etiologic definition of radon-related risk: the probability of dying *prematurely* from a radon-induced lung cancer. The difference is that the BEIR VI method omits that proportion of radon-related lung cancer mortality occurring in individuals who would have died later from lung cancer in the absence of radon exposure. BEIR VI presents estimates of "attributable risk," by which was meant the difference between the lung cancer mortality in an exposed and unexposed population, divided by the mortality in the exposed population; in contrast, EPA here presents estimates of the "etiologic fraction" (*EF*), which represents the fraction of lung cancer deaths in the exposed population in which radon played some causative role.

3) EPA adds to the discussion found in BEIR VI on how changes in smoking patterns might impact estimates of risk. It will be shown that estimates of *EF* are much less sensitive to changes in smoking prevalence than are estimates of risk per WLM.

Section B details life-table methods for deriving lifetime risks. We present results for *EF*, risk per WLM, and *YLL* per cancer death in Sections C through E. Section F compares current estimates to the previous EPA estimates. Section G discusses health risks other than lung cancer mortality. Section H considers the problem of estimating radon-induced lung cancer deaths among current smokers. Section I offers a discussion on estimation problems related to smoking. A very short summary is given in Section J.

B. Life-Table Derivation of Lifetime Risks of Radon-Induced Lung Cancer Death

Lifetime risk estimates such as risk per WLM can be derived using a life-table method. Life-table methods account for the effects of competing causes of death, which is necessary because the probability of dying from a radon-induced lung cancer depends on the age-specific rates of death from all causes as well as lung cancer death rates. The death rates from lung cancer and from all causes are determined from U.S. vital statistics. The risk per WLM and *EF* estimates are calculated assuming stationary populations for male ES, female ES, male NS, and female NS. This results in risk estimates for each of these four stationary populations.

Calculating risk per WLM is essentially a four-step process. First, the age-specific (baseline) lung cancer death rates are determined for each of the four stationary populations. As described in detail below, the rates are derived from the vital statistics on lung cancer death, recent data on ever-smoking prevalence, and by assuming that the ES age-specific lung cancer rates are 14 (males) or 12 (females) times higher than the rates for NS. Second, a model for age-specific relative risks is chosen and applied to the baseline rates to determine the age-specific lung cancer risk due to a constant, lifelong radon exposure. The third step is to calculate a weighted-average of these age-specific risks using weights equal to the probability of survival (to each age). This step is used to yield separate risk per WLM estimates for the four gender- and smoking-specific populations. The final step combines these estimates to obtain the risk per WLM for the entire U.S. population. Details on each of these steps follow.

1. Lung cancer death rates for male and female ES and NS: Baseline lung cancer death rates for the general population are derived from 1989-91 vital statistics (NCHS 1992, 1993a, 1993b). To obtain the lung cancer death rates for ES and NS, we assume, as in BEIR VI, that the lung cancer death rates are 14 times (males) or 12 times (females) greater for ES than NS, independent of age. The lung cancer death rates are then calculated from the age-specific proportions of ES in the general population, as is shown below. First, note that

$$h_{pop}(x) = (1-p(x))\, h_{NS}(x) + p(x) \cdot h_{ES}(x),$$

where $h_{NS}(x)$, $h_{ES}(x)$ and $h_{pop}(x)$ are the respective lung cancer death rates for NS, ES, and the general population, and $p(x)$ is the proportion of ES at age x. Letting RR denote the smoking related relative risk (14 for males, 12 for females), and substituting for $h_{ES}(x)$ yields:

$$h_{pop}(x) = (1-p(x))\, h_{NS}(x) + p(x) \cdot RR \cdot h_{NS}(x),$$

or equivalently:

$$h_{NS}(x) = h_{pop}(x)\, [(1-p(x)) + p(x) \cdot RR]^{-1}, \tag{4a}$$

We used Equation 4a to calculate the rates for NS, and Equation 4b for ES:

$$h_{ES}(x) = RR \cdot h_{NS}(x) \qquad\qquad (4b)$$

In BEIR VI, it was assumed that 58% of males and 42% of females are ES, for all ages $\geq$ 18 y. As an illustration, consider how this formula would be applied for males of age 70 y. In the U.S., the lung cancer death rate, $h_{pop}(70)$ for such males was 0.0044. If 58% were ES, the corresponding rates for NS and ES would be, according to Equations 4a and 4b:

$$0.000515 = 0.0044 \, [0.42 + 14(0.58)]^{-1} \quad \text{for NS}$$
$$0.0072 \quad = 14 \times 0.000515 \qquad\qquad \text{for ES}$$

Extending the basic approach in BEIR VI, we allow ES prevalence to depend on age. Estimates of smoking prevalence in 1990 for males and females, shown in Figure 1, are based on data from six NHIS surveys (DHHS 1997). Details are given in Appendix A. Figure 1 clearly indicates that for the cancer-prone ages between 50 and 80 y, the male ever-smoking prevalence substantially exceeded 58%. As a result, our corresponding estimates of NS and ES male lung cancer death rates for these critical ages are somewhat smaller than in BEIR VI. For example, our estimate of ever-smoking prevalence for males of age 70 y is 74%. Applying this prevalence to Equations 4a and 4b yields the NS and ES lung cancer death rates:

$$0.000414 = 0.0044 \, [0.26 + 14(0.74)]^{-1} \quad \text{for NS}$$
$$0.0058 \quad = 14 \times 0.000414 \qquad\qquad \text{for ES}$$

These rates are about 21% smaller than the BEIR VI rates, which were based on the assumption that prevalence rates are age-independent.

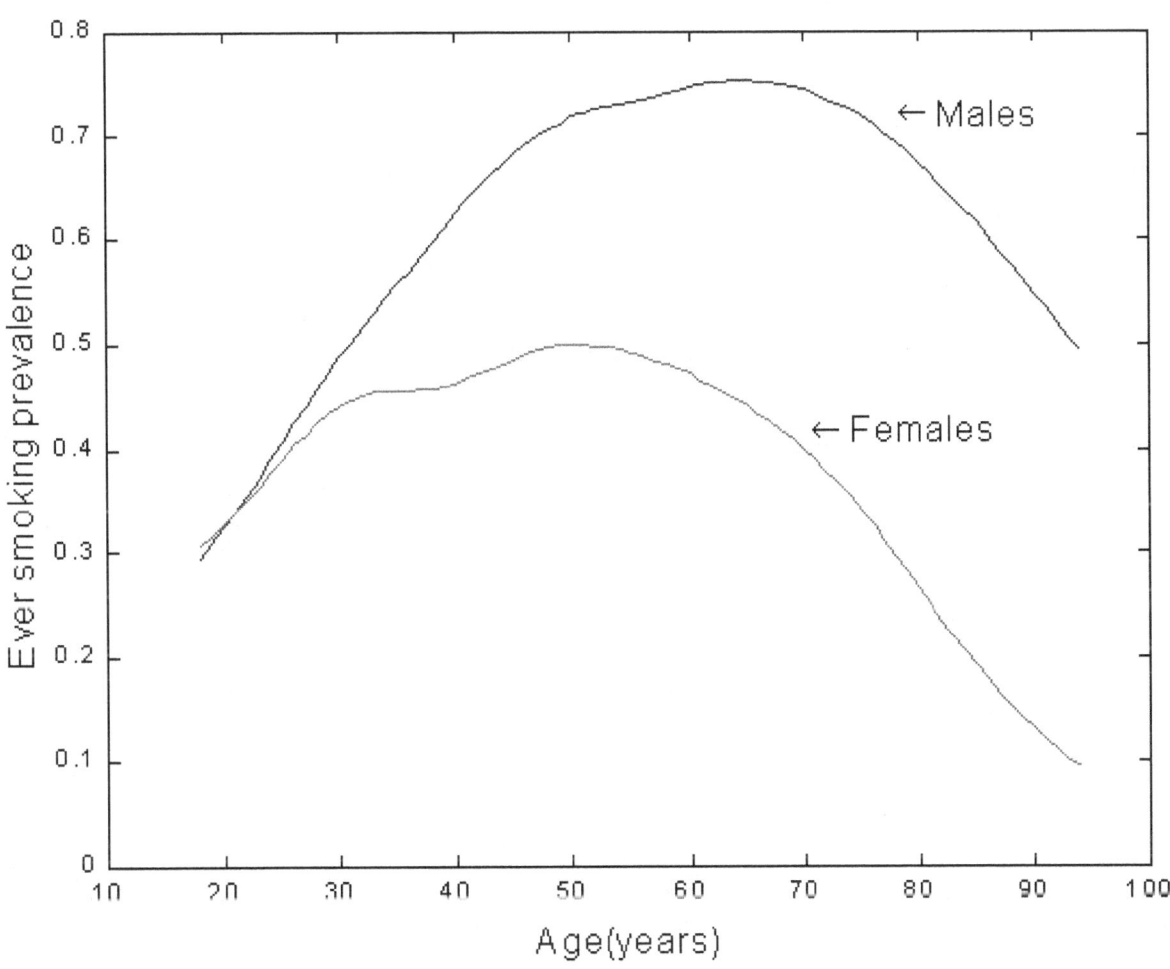

Figure 1: Ever smoking prevalence by age and gender.

2. Choice of a relative risk model: As in BEIR VI, radon-induced lung cancer death rates were obtained simply as a product of the modeled age-specific excess relative risks, $ERR(x)$, and the baseline lung cancer rate from all causes, $h(x)$. For modeling the relative risks we used a scaled version of the concentration model, one of the two models preferred by the BEIR VI committee. The scaling results in lifetime risk estimates intermediate between results that would be obtained from the BEIR VI concentration and duration models.

The concentration model assumes that the risk per unit exposure increases as the radon decay product concentration (*i.e.*, the exposure rate) decreases down to some limiting value, whereas the duration model assumes that the risk increases as the exposure duration is increased to some limiting value. Obviously these two approaches are closely related, since, for fixed total exposure, increased duration means decreased exposure rate. Under some exposure conditions the two approaches are essentially equivalent, and the BEIR VI committee found that the two models fit the miner data equally well.

One might try to select one of these two models on the basis of biological plausibility. An inverse dose rate effect has been seen in cellular studies of alpha-particle induced mutations and transformation. If one postulates that the carcinogenic action of radon stems from the mutagenicity of alpha radiation, the critical factor in determining the risk per unit exposure would be the exposure rate (concentration), and only secondarily, the duration. On the other hand, the potency of a promoter may depend directly on exposure duration, as well as concentration. It turns out, however, that one cannot distinguish the two models on this basis because the BEIR VI analysis was carried out on highly averaged data, not reflective of the day-to-day, or even the year-to-year, variations in concentrations to which miners were exposed to. Moreover, the categorization of exposure rates and exposure durations are somewhat arbitrary, and these categorizations may have had some effect on the limiting value for the risk per unit exposure projected with each of the models. Thus, the difference in risk projections from the two models may be largely an artifact of the analysis, and neither projection has more credibility than the other. Therefore, to arrive at a "best estimate" of risk, it is reasonable to average the two models in some way.

One approach would be for EPA to calculate risk with both models, on a case-by-case basis, and average the results. This would be cumbersome, and it is preferable to have a single model for calculating risks. Since the two models recommended in BEIR VI exhibit very similar dependencies on age and time-since-exposure (see Table 3), as well as the same two-fold higher risk for never smokers, a simple approach to averaging is to adjust one of the models in such a way as to yield results approximately midway between those obtained using the two unmodified models.

We chose to modify the concentration model for this purpose because, as will be shown in the next section, the concentration model avoids ambiguities that may arise when assessing health impacts from residential exposures at levels that change over time. As shown in Table 6, the risk per WLM is 6.52×10^{-4} for the concentration model and 4.43×10^{-4} for the duration model. We scaled the concentration model so that the risk per WLM would equal the geometric mean of these two values (5.38×10^{-4}). This is easily achieved since (see Section VI.B.4), the risk per WLM is approximately proportional to the risk coefficient β. The risk coefficient for the EPA's model (scaled-concentration model) is:

$$\beta = 0.0768 \times (4.43 / 6.52)^{1/2} = 0.0634, \tag{5}$$

and the risk per WLM is $5.38 \times 10^{-4} \approx (6.52 \times 10^{-4}) \times (4.43 / 6.52)^{1/2}$.

Table 6: Risk per WLM based on BEIR VI concentration and duration models

Model	Risk per WLM (10^{-4})
Concentration	6.52
Duration	4.43

Details on how the concentration and duration models were applied to obtain the values in Table 6 are given in the next section.

3. Applying the concentration and duration models: As described in Part IV, the BEIR VI concentration model specifies that the excess relative risk (relative risk -1) depends on time-since-exposure, attained-age, and rate of exposure (concentration) according to the formula:

$$ERR = \beta \, (w_{5\text{-}14} + \theta_{15\text{-}24} \, w_{15\text{-}24} + \theta_{25+} \, w_{25+}) \, \phi_{age} \, \gamma_z, \tag{3}$$

The θ-parameters detail how relative risk depends on time-since-exposure, and ϕ_{age} describes the dependency on attained age. The γ_z, ranging from 1 for radon concentrations below 0.5 WL to 0.11 for concentrations above 15 WL, define the dependency on exposure rate. This formula can be simplified by noting that γ_z is almost always equal to 1, because residential exposure rates are almost always below 0.5 WL. Letting $\beta^* = \beta \, \phi_{age}$, and using the (unadjusted) parameter estimates from BEIR VI given in Table 3, the formula for the excess relative risk may then be expressed as.

23

$$ERR = \beta^* (w_{5\text{-}14} + 0.78\ w_{15\text{-}24} + 0.51\ w_{25+}),$$

where $\beta^* = 0.0768$ for attained age $(x) < 55$ y

$\qquad\qquad = 0.0438$ for $55\ y \leq x < 65$ y

$\qquad\qquad = 0.0223$ for $65\ y \leq x < 75$ y

$\qquad\qquad = 0.0069$ for $x \geq 75$ y.

This formula might be applied, for example, to estimate health effects at age 60 y from a residential radon exposure at level 6 pCi/L (0.867 WLM/y) up to age 45 y, and 2pCi/L (0.289WLM/y) for the next 15 y. The estimated proportional increase in the risk of a fatal lung cancer at age 60 y would be about 110%:

$$1.10 = 0.0438[2.89+0.78(8.67)+0.51(30.35)]$$

Figure 2 shows how the modeled excess relative risks for a constant lifetime exposure depend on attained age. Up to age 55, the relative risks increase because cumulative (weighted) exposures increase with age. The excess relative risks then drop (discontinuously) at ages 55, 65, and 75. To avoid such biologically implausible discontinuities, we use splines to smooth this function (see Figure 2) for our calculations. The excess relative risk function is then multiplied by baseline rates to yield age-specific rates of radon-induced lung cancer death. These were then averaged as described in the next sections to yield the estimate of 6.52 deaths per 10,000 WLM in Table 6.

We now turn our attention to the duration model. For constant exposures and attained ages greater than 35 y, the duration model (and simple algebra) simplifies to:

$$ERR = \beta^*(w_{5\text{-}14} + 0.72\ w_{15\text{-}24} + 0.44\ w_{25+}),$$

where now $\quad \beta^* = 0.0561$ for attained age $(x) < 55$ y

$\qquad\qquad = 0.0292$ for $55\ y \leq x < 65$ y

$\qquad\qquad = 0.0157$ for $65\ y \leq x < 75$ y

$\qquad\qquad = 0.0073$ for $x \geq 75$ y.

Unfortunately, the duration model does not adequately specify how to calculate risks that result from exposures with changing radon levels. Returning to the example in which the residential exposure level changes at age 45 y, it is not clear whether the risk from radon exposures received by age 55 y should be calculated as the sum of risks from two separate exposures of duration 35 y ($\gamma_z = 10.2$) and 10 y ($\gamma_z = 2.78$), or whether the appropriate duration is 45 y ($\gamma_z = 10.2$).

Figure 3 shows how the ERR's from a constant lifetime exposure depend on attained age for smoothed versions of the duration and scaled concentration models (see also Appendix B for details). Figures 4 and 5 show age-specific estimates of lung cancer death rates for male ES, male NS, female ES and female NS. Estimates of rates

24

from exposure to radon were derived using the scaled concentration model. Not surprisingly, radon-related rates of lung cancer death are many times higher for ES than NS. Figure 6 shows the lung cancer death rates for a stationary population that comprises all four subpopulations. These rates are weighted averages of the four sets of age-specific rates. Formulas for averaging death rates and survival functions from different populations are given in Appendix C.

Caution is warranted in interpreting the lung cancer death rates shown in Figures 4 - 6, especially those linked to exposure to radon. One would infer from those figures that whereas lung cancer death rates from all causes would increase consistently from age 40 y to about age 85 y, the rates of premature lung cancer death due to exposure to radon are greatest between ages 55 y and 75 y. However, the precise form of the temporal dependence of the risk is less certain than the estimate of lifetime risk. This is because estimates of lifetime risk are determined using the mortality experience of miners at all ages, whereas age-specific estimates are largely determined by the miners' mortality experience in restricted age intervals.

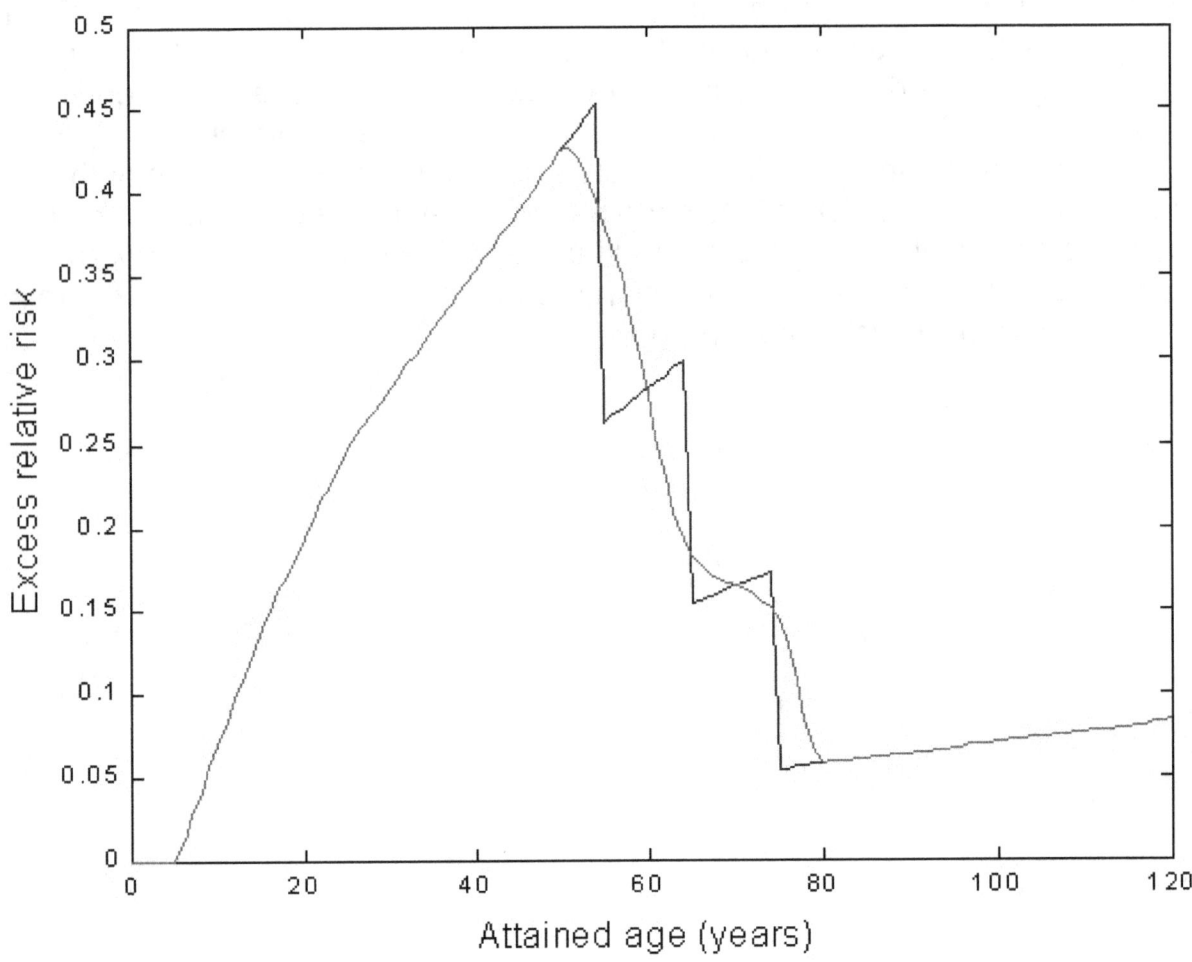

Figure 2: BEIR VI (unscaled) concentration model age-specific excess risks from a 0.181 WLM/y radon exposure. Smoothed version also shown.

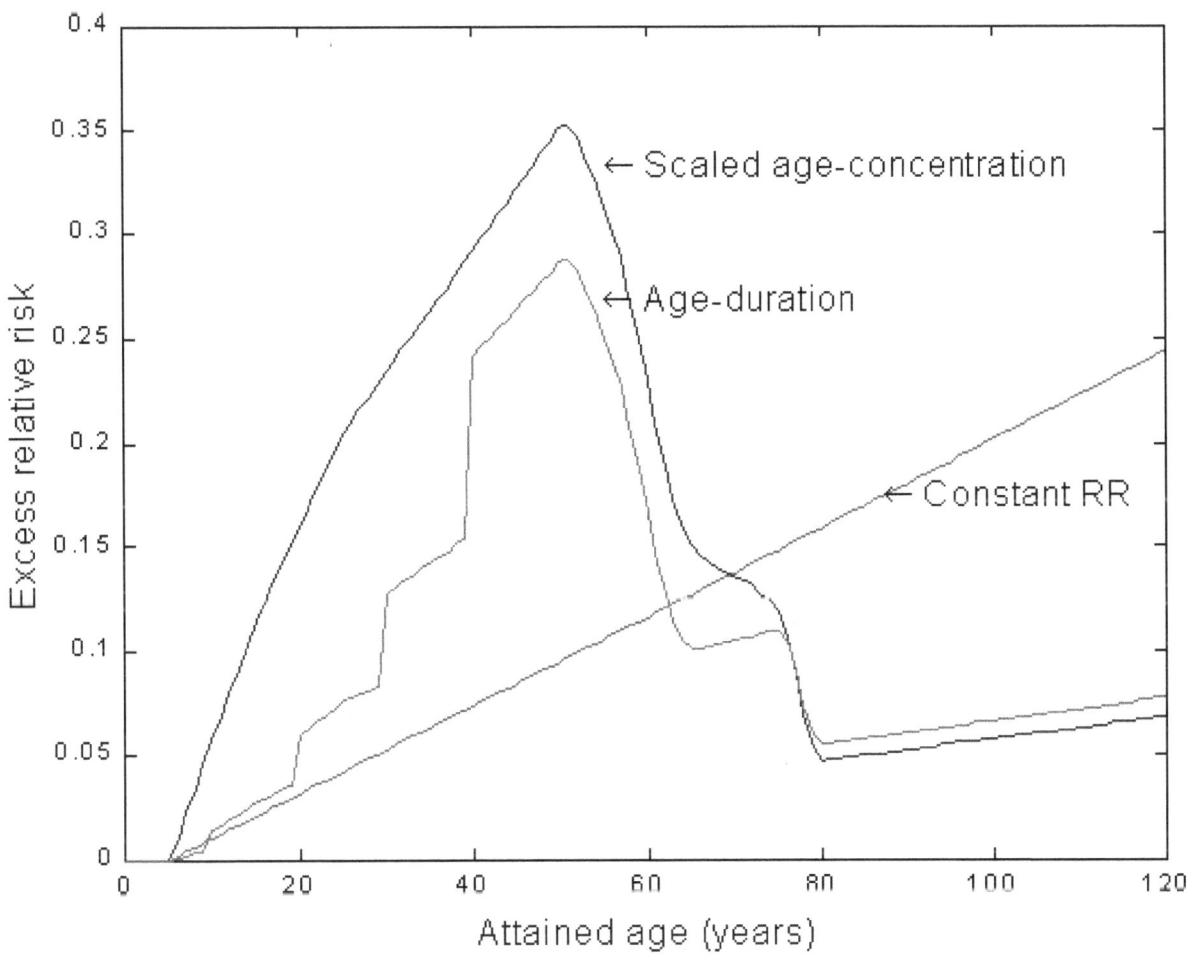

Figure 3: Smoothed age-specific excess relative risks from a constant radon exposure at rate 0.181 WLM/y.

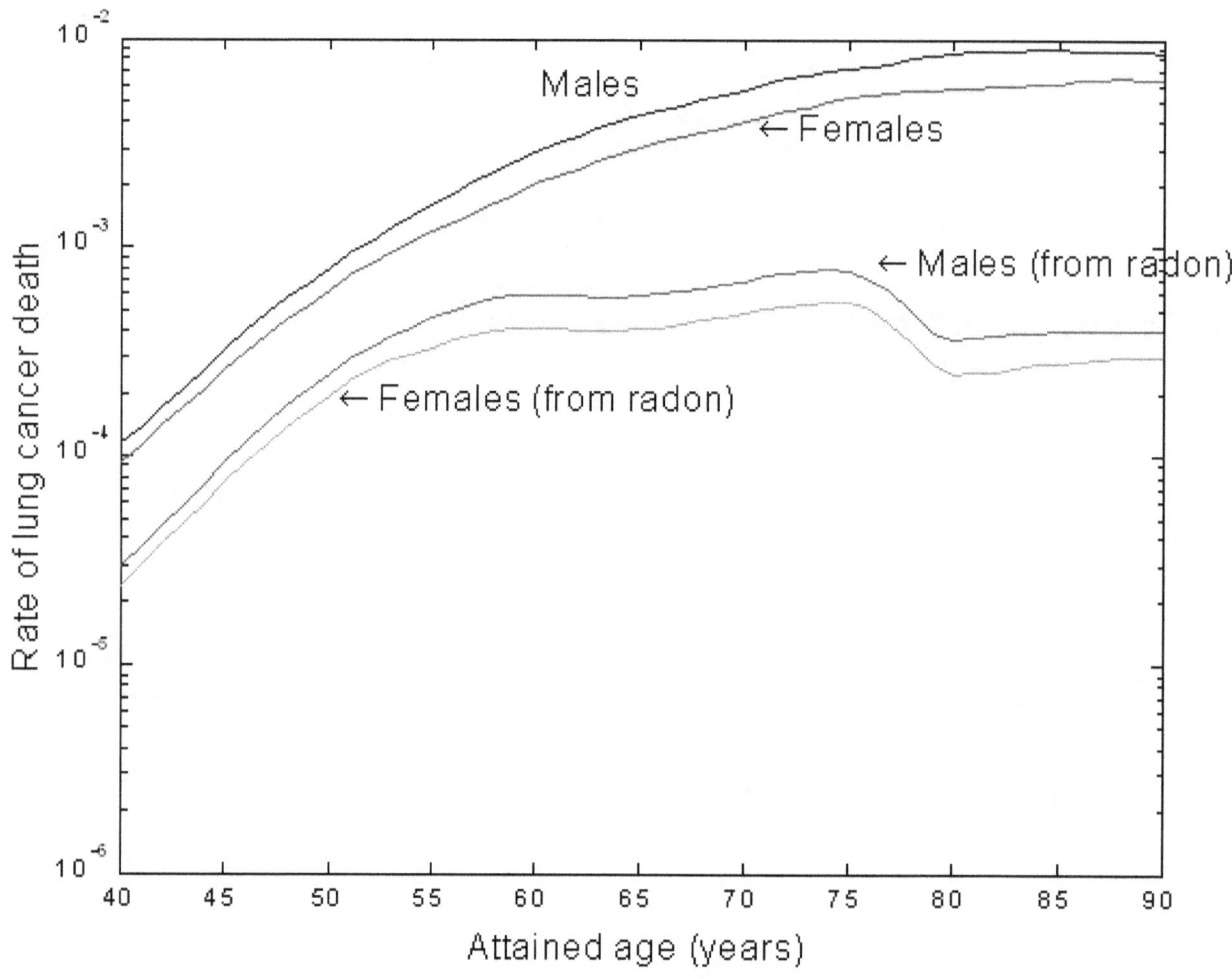

Figure 4: Rates of lung cancer death for ES males and females. Estimated rates of premature lung cancer death due to a constant exposure to radon of 0.181 WLM/y are also shown. See the text for a discussion of uncertainties associated with these estimates.

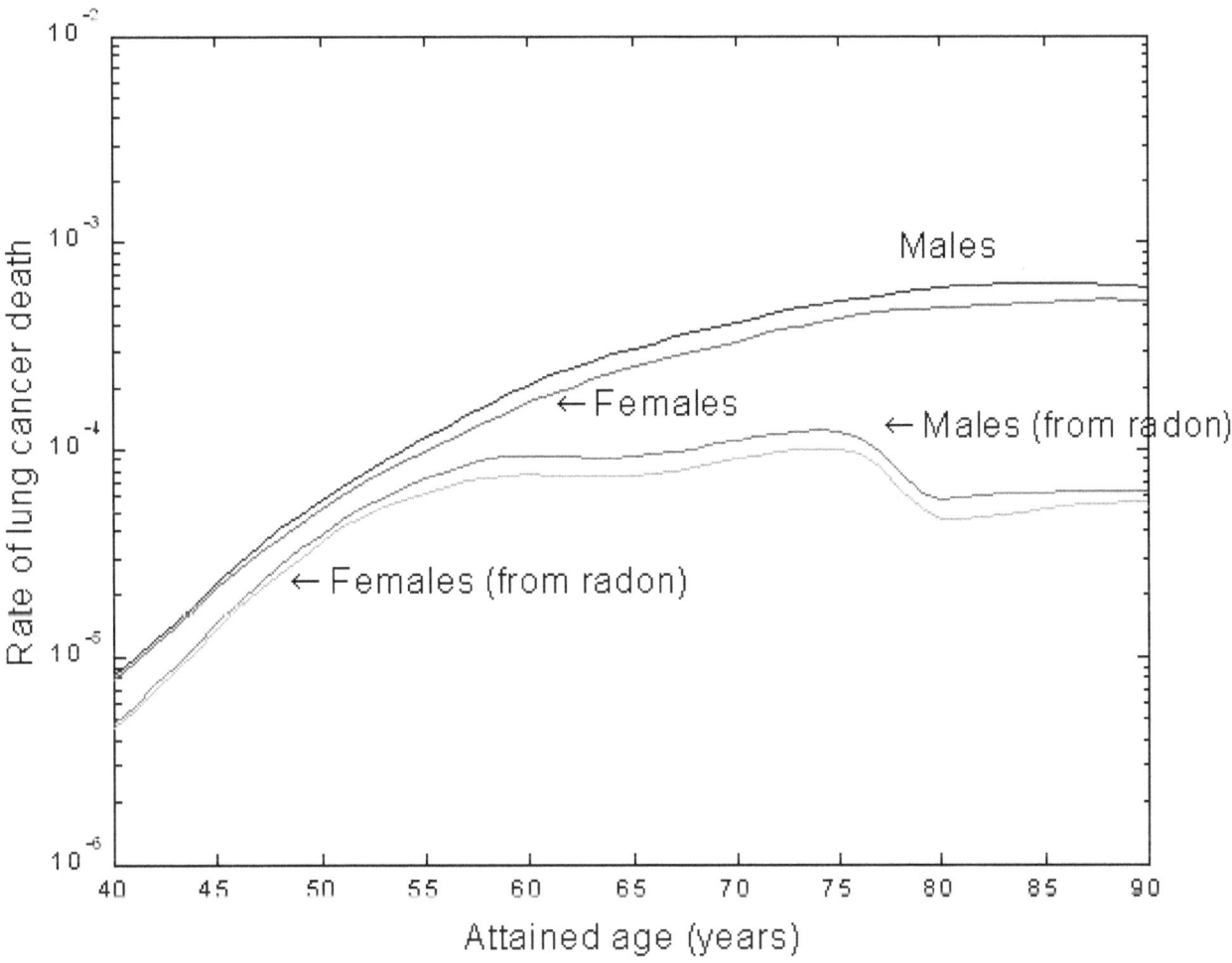

Figure 5: Rates of lung cancer death for NS males and females. Estimated rates of premature lung cancer death due to a constant radon exposure at rate 0.181 WLM/y also shown. See the text for discussion of uncertainties associated with these estimates.

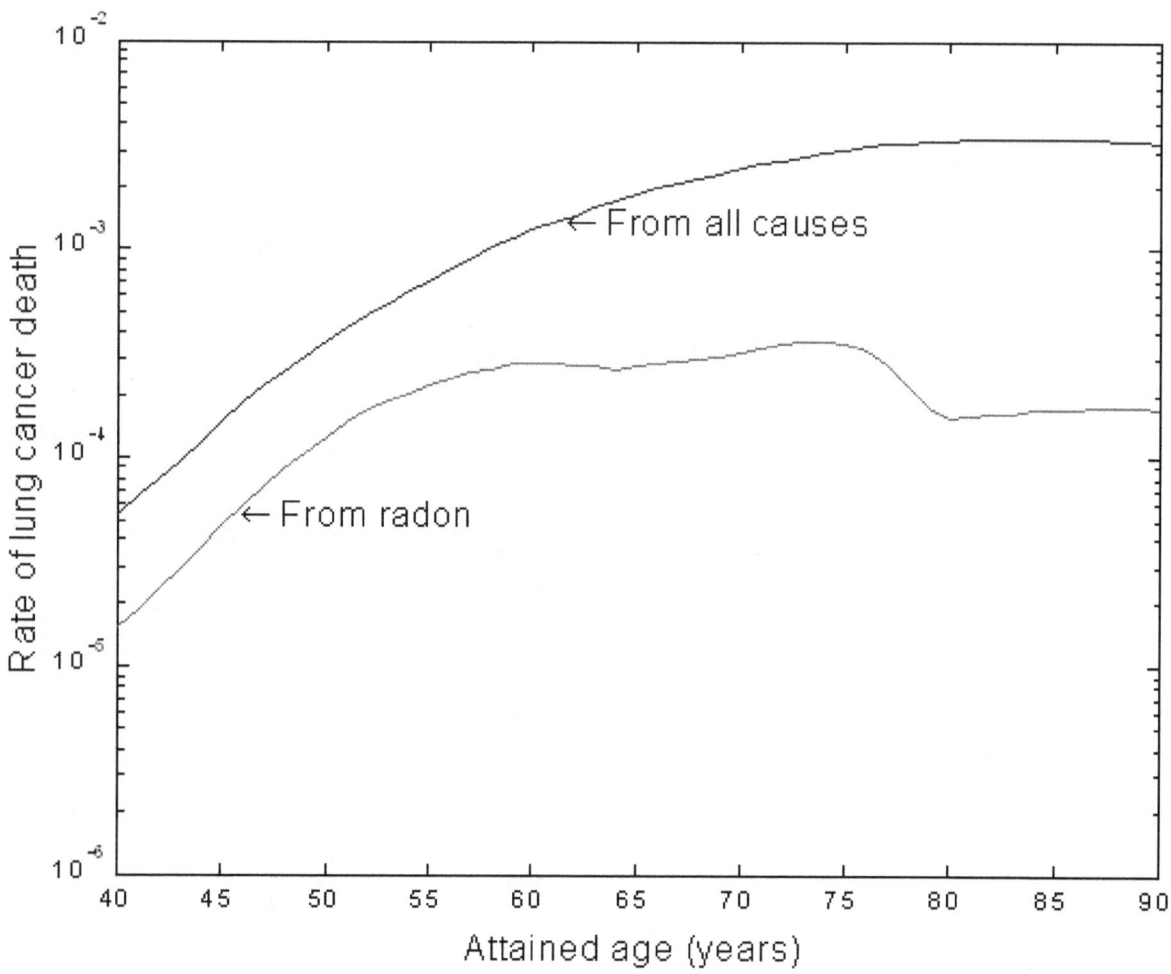

Figure 6: Rates of lung cancer deaths for a stationary population in which 53% of males and 41% of females are ES. Rates of premature lung cancer death due to a constant radon exposure at rate 0.181 WLM/y also shown. See the text for discussion of uncertainties.

4. Averaging the age-specific risks of lung cancer death: Weighted averages of the age-specific excess lung cancer death rates shown in Figures 4 and 5 are calculated to yield the risk estimates for male and female ES and NS. The weights are the probabilities of survival, and the averaging is accomplished through integration. Details follow.

Let $S(x)$ be the probability of survival to age x for one of the gender- and smoking-specific stationary populations, and assume a constant excess residential radon exposure rate (WLM/y) equal to Δ. The survival function accounts for the increased probability of smoking-related lung cancer death, but, for reasons discussed in Section I below, is not adjusted for smoking-related risks other than lung cancer. Let $S(x, \Delta)$ be the probability of survival to age x, adjusted to account for a small incremental lifetime excess rate of radon exposure equal to Δ (for our calculation we used $\Delta = 0.00181$ WLM/y). Also, let $h(x)$ be the baseline lung cancer death rate, adjusted to account for effects of smoking, and $e(x, \Delta)$ be the ERR at age x due to the excess exposure (at a rate $= \Delta$). The formula for the lifetime risk per WLM ($RWLM$), is:

$$RWLM = \frac{\int_0^\infty h(x) \cdot e(x, \Delta) \cdot S(x, \Delta) \cdot dx}{\int_0^\infty \Delta \cdot S(x, \Delta) \cdot dx}$$

The formula for calculating lifetime etiologic fraction (EF) is similar. The EF is the risk of a premature lung cancer death from the background exposure of $g_b(x)$ (measured in WLM/y) divided by the baseline lifetime risk of lung cancer death from all causes. (See Greenland and Robins (1998) for an interesting discussion of problems associated with estimating the etiologic fraction). A formula for the risk (R) of a premature cancer death due to radon is:

$$R = \int_0^\infty h(x) \cdot e(x, g_b(x)) \cdot S(x, g_b(x)) \cdot dx$$

However, for constant $g_b(x) = g_0$, the following linear approximation for R holds:

$$R \approx (g_0 / \Delta) \cdot \int_0^\infty h(x) \cdot e(x, \Delta) \cdot S(x, \Delta) \cdot dx$$

The formula for the baseline risk is:

$$R_{baseline} = \int_0^\infty h(x) \cdot S(x) \cdot dx$$

Our estimate of *EF* from an exposure of 0.181 WLM/y is:

$$EF = \frac{R}{R_{baseline}} = \frac{(0.181/\Delta) \cdot \int_0^\infty h(x) \cdot e(x,\Delta) \cdot S(x,\Delta) \cdot dx}{\int_0^\infty h(x)S(x) \cdot dx}$$

The average years of life lost per radon-induced lung cancer death (*YLL*) is obtained through:

$$YLL = \frac{\int_0^\infty (S(x) - S(x,\Delta)) \cdot dx}{\int_0^\infty h(x) \cdot e(x,\Delta) \cdot S(x,\Delta) \cdot dx}$$

   5. Combined risk estimates for the U.S. population: A combined risk per WLM estimate for the entire population is calculated as a weighted average of the male ES, male NS, female ES, and female NS risks. The weights are proportional to the expected number of person-years for each gender-and-smoking category. Similarly, the combined *EF* and combined *YLL* estimates are weighted averages of the corresponding gender- and smoking-specific estimates. For *EF*, the weights are proportional to the lifetime baseline cancer death probabilities. For *YLL*, the weights are proportional to the lifetime risks of a radon-induced lung cancer death. Details are given in Appendix C.

C. Etiologic Fraction

Table 7 shows estimates for the *EF*, or the proportion of lung cancer deaths induced by radon, for male and female ES and NS. These estimates have been calculated using life-table methods applied to the BEIR VI age-concentration model as detailed in Section B. We assumed a constant rate of radon exposure of 0.181 WLM per year, as detailed in Section F. The estimates indicate that radon exposure accounts for about 1 in 8 ES lung cancer deaths and about 1 in 4 NS lung cancer deaths. These estimates are subject to uncertainties, which are quantified when feasible in Section VIII.E. For example, 90% uncertainty bounds calculated for the ES suggest that the *EF* for this group is between 0.05 and 0.3, or that the estimates shown in Table 7 for ES may be accurate within a factor of about 3. Estimates for NS would be subject to greater uncertainties since most of the miners were ES.

Table 7: Estimated etiologic fraction[a] by smoking category and gender.

Gender	Smoking Category	
	ES	NS
Male	0.129	0.279
Female	0.116	0.252

[a] Based on 1989-91 vital statistics and mortality data (NCHS 1992, 1993a, 1993b, 1997). See the text for a discussion of uncertainties.

The *EF* estimates in Table 7 for male and female ES and NS have been multiplied by the corresponding estimates, shown in Table 8, of the lung cancer deaths in 1995 (NAS 1998). The result of these calculations are estimates of the lung cancer deaths due to radon progeny for male and female ES and NS. The calculated total number of radon-induced lung cancer deaths in 1995 was about 21,100: 13,000 males and 8,100 females; 18,200 ES and 2,900 NS. The uncertainties in these estimates are quantified in Section VII.E.

Table 8: Estimated fraction of lung cancer deaths in 1995 attributable to radon.

Gender	Smoking Category	Number of Lung Cancer Deaths in 1995	Fraction Due to Radon[a]	Number of Radon-induced Deaths in 1995
	ES	90,600	0.129	11,700
Male	NS	4,800	0.279	1,300
	ES and NS	95,400	0.136	13,000
	ES	55,800	0.116	6,500
Female	NS	6,200	0.252	1,600
	ES and NS	62,000	0.131	8,100
	ES	146,400	0.124	18,200
Male & Female	NS	11,000	0.263	2,900
	ES and NS	157,400	0.134	21,100

[a] Estimates of the fraction due to radon are subject to uncertainties as discussed in the text.

An estimated 13.4% of lung cancer deaths in 1995 were radon-related. This percentage depended on the proportion of ES among adults of lung-cancer prone ages, because (see Table 7) etiologic fractions are about 2 times greater for NS than ES. Theoretically, this *EF* could change, because the *EF* depends on the age-specific ES prevalences, and these prevalences change. For example, the proportion of male ES is much greater for people of lung cancer prone ages in 1995 than for the entire male adult population. Over 70% of males between ages 50 and 80 years are ES compared to an average of 58% for all adult males. It also seems likely that the proportion of children less than 18 y who will take up smoking will be considerably lower than the ES proportion among adults.

To calculate the *EF* for all living males and females, we first assume that 37% of males and 36% of females of ages less than 18 y will be ES. These percentages are derived by noting that, for the youngest cohort (born 1965 to 1969) for which we have reliable data, the proportion of ES in 1990 was about 37% for males and 36% for females (DHHS 1997). It then follows that, since the ES prevalence for adults is about 58% for males and 42% for females, and about 27% of males and 24% of females are of ages less than 18 y, about 53% of living males and 41% of females would be ES. We can then use life-table calculations based on a stationary population for which the same 53% of males and 41% of females would be ES. Results are given in Table 9.

34

Table 9: Estimated etiologic fraction by smoking category and gender for a stationary population in which 53% of males and 41% of females are ES.

Gender	Smoking Category	Etiologic Fraction[a]
Male	ES and NS	0.139
Female	ES and NS	0.132
Male and Female	ES	0.124
Male and Female	NS	0.264
Male and Female	ES and NS	0.136

[a] Based on 1990 adult (ages ≥18 y) ever-smoking prevalence data (58.7% males and 42.3% females are ES) and assumption that 37% (males) and 36% (females) of children (ages < 18 y) will become ES.

Of course we do not know what percentage of children will become smokers. Results of national surveys indicate that smoking prevalence among high school students has increased since 1990, and is only recently showing signs of leveling off. Current smoking estimates in youth, defined as tobacco use on at least one of the last 30 days, has increased from about 28% in 1991 to about 36% in 1997 (Bergen and Caporaso 1999) and 1998 (CDC 2000).

The striking similarity of the *EF* estimates in Tables 8 and 9 reflect the fact that the *EF* is generally insensitive to changes in smoking prevalence. This is true unless the ES prevalence is very small. As shown in Figure 7, the *EF* decreases relatively rapidly until the ES prevalence is about 0.2, and then gradually flattens out. For ES prevalence between 0.2 and 1.0, the *EF* decreases from about 0.16 to 0.13 for males, and 0.15 to 0.12 for females. The next section shows that, in contrast to the *EF*, the risk per WLM is sensitive to changes in ES prevalence.

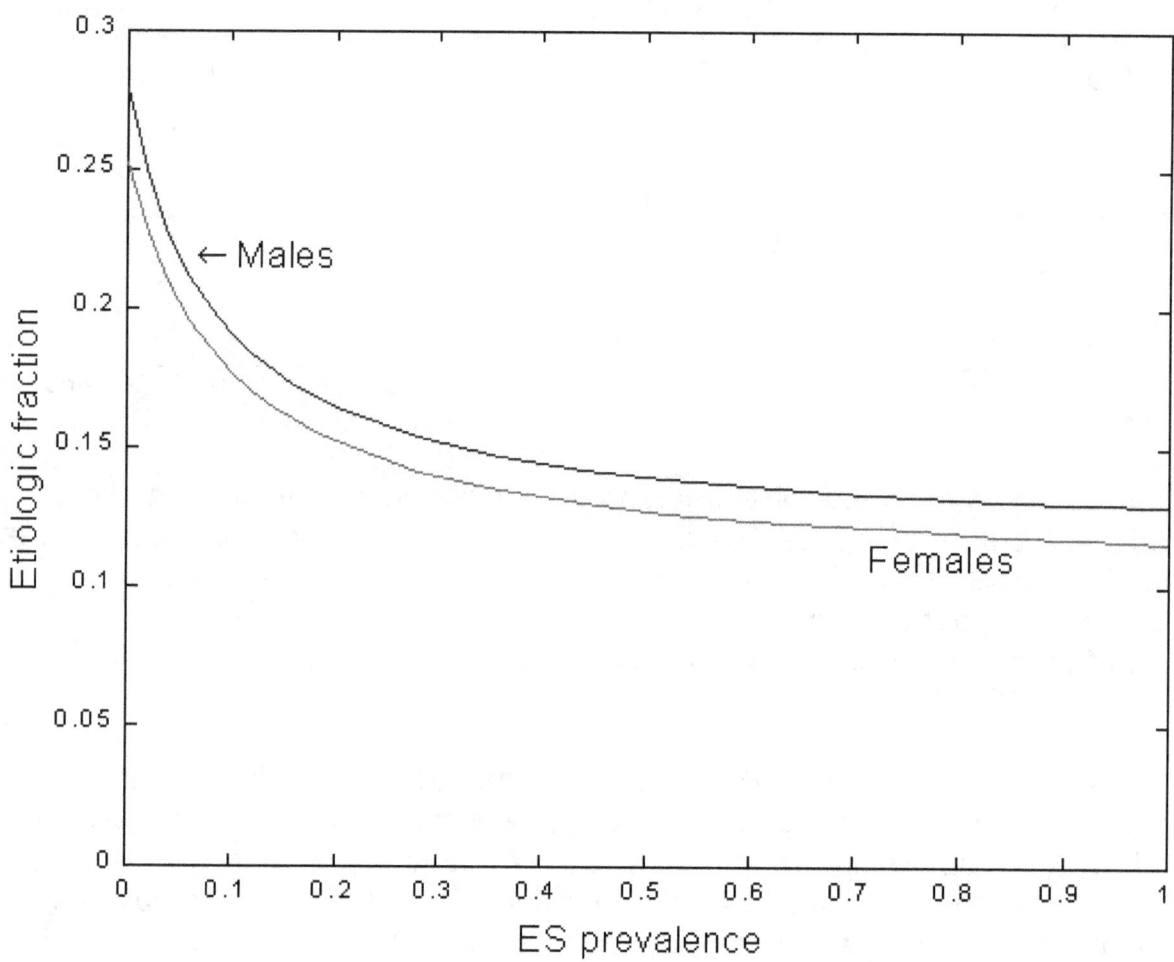

Figure 7: Etiologic fraction by ES prevalence from a lifetime exposure of 0.181 WLM/y

D. Risks per Unit Exposure and per Unit Concentration

Table 10 presents estimates of risk per WLM by smoking category and gender. These estimate the number of expected radon-induced cancers for the current population divided by the corresponding total of past and expected future radon exposures. The estimates have been derived using life-table methods assuming, as in BEIR VI, that radon exposure rate is constant during the life of each individual. Risk estimates for NS and ES have been combined by assuming a stationary population for which 53% of males and 41% of females would be ES. For the entire population, the risk estimate is 5.38×10^{-4} fatal lung cancers per WLM. This estimate is subject to uncertainties, as described in Chapter VII. Ninety percent uncertainty interval for the risk per WLM ranges from 2×10^{-4} and 12×10^{-4}.

The estimated risks for ES and NS are, respectively, about 1.8 and 0.3 times that for the general population. Thus, ES are estimated to have about 6 times the risk from radon as NS. Figure 8 shows that the estimated risk per WLM is sensitive to changes in ES prevalence. Our risk estimates are based on the premises that 36% - 37% of children will smoke and that children make up about a quarter of the population. Calculations show that the proportion of those now alive who would smoke sometime during their lifetime would be about 53% for males and 41% for females. On the other hand, if all children were to remain never smokers, the corresponding risk per WLM estimates would be about 15% lower than those given in Table 10. Besides ES prevalence, the baseline lung cancer rates, and thus also risk per WLM, will be affected by other changes in smoking patterns, including quit rates and number of cigarettes smoked.

What do these risk estimates mean for homeowners who have had the radon level in their home measured? Such measurements are usually given in picocuries per liter (pCi/L) of radon gas. Assuming that, on average, people spend about 70% of their time indoors at home and that the equilibrium fraction between radon and its daughters is 40% (NAS 1999), it follows from the definitions in Section II that, at 1 pCi/L of radon gas, the radon daughter exposure rate is 0.144 WLM per year.

From Table 10, the average risk of a fatal lung cancer due to lifetime exposure at 1 pCi/L is then:

$$(0.144 \text{ WLM/y}) (75.4 \text{ y/lifetime}) (5.38 \times 10^{-4} / \text{WLM}) = 0.58\%$$

In general, if the concentration in pCi/l is C, the estimated risk from lifetime exposure will be $0.0058 \, C$. Hence lifetime exposure at the EPA action level of 4 pCi/l corresponds to an estimated risk of 2.3%. Similarly, for ES and NS, the lifetime risks are $0.0103 \, C$ and $0.0018 \, C$, respectively. Again, risks for ES are almost 6 times higher than for NS. Risks for current smokers would likely be higher than for ES. Estimates of lifetime risks for NS and current smokers at constant concentrations are tabulated in Appendix D.

Figure 9 provides information on how risks may depend on age at exposure. Plotted there is the calculated lifetime risk per WLM as a function of age at exposure for "average" members of the population. Results are also shown for ES and NS. We can apply the results shown in Figure 9 to approximate risks for specific exposure intervals. For example, consider an individual exposed to radon at 1 pCi/L between ages 40 y and 41 y. The estimated risk per WLM for exposures received between ages 40 y and 41 y is 7.71×10^{-4} WLM^{-1} so the risk for such an exposure would be about:

$$(0.144 \text{ WLM y}^{-1})(1 \text{ y})(7.71 \times 10^{-4} \text{ WLM}^{-1}) = 0.011\%$$

Calculations for ES and NS, or for newborns destined to be ES or NS, would be done in a similar manner. Again, on an age-specific basis, the model estimates for ES and NS are approximately 180% or 30% of the general population estimate, respectively. Caution must also be applied for assessing risks because of uncertainties associated with age-specific risks, particularly with childhood exposures (see: Sections VII.C.3 and VII.C.4).

Table 10: Estimates of risk per WLM by smoking category and gender for a stationary population in which 53% of males and 41% of females are ES.

Gender	Smoking Category	Risk per WLM[a] (10^{-4})	Expected Life Span[a] (years)
Male	ES	10.6	71.5
	NS	1.74	72.8
	ES and NS	6.40	72.1
Female	ES	8.51	78.0
	NS	1.61	79.4
	ES and NS	4.39	78.8
Male & Female	ES	9.68	74.2
	NS	1.67	76.4
	ES and NS	5.38	75.4

[a] Based on 1990 adult (ages ≥18 y) ever-smoking prevalence data (58.7% males and 42.3% females are ES) and assumption that 37% (males) and 36% (females) of children (ages < 18 y) will become ES.

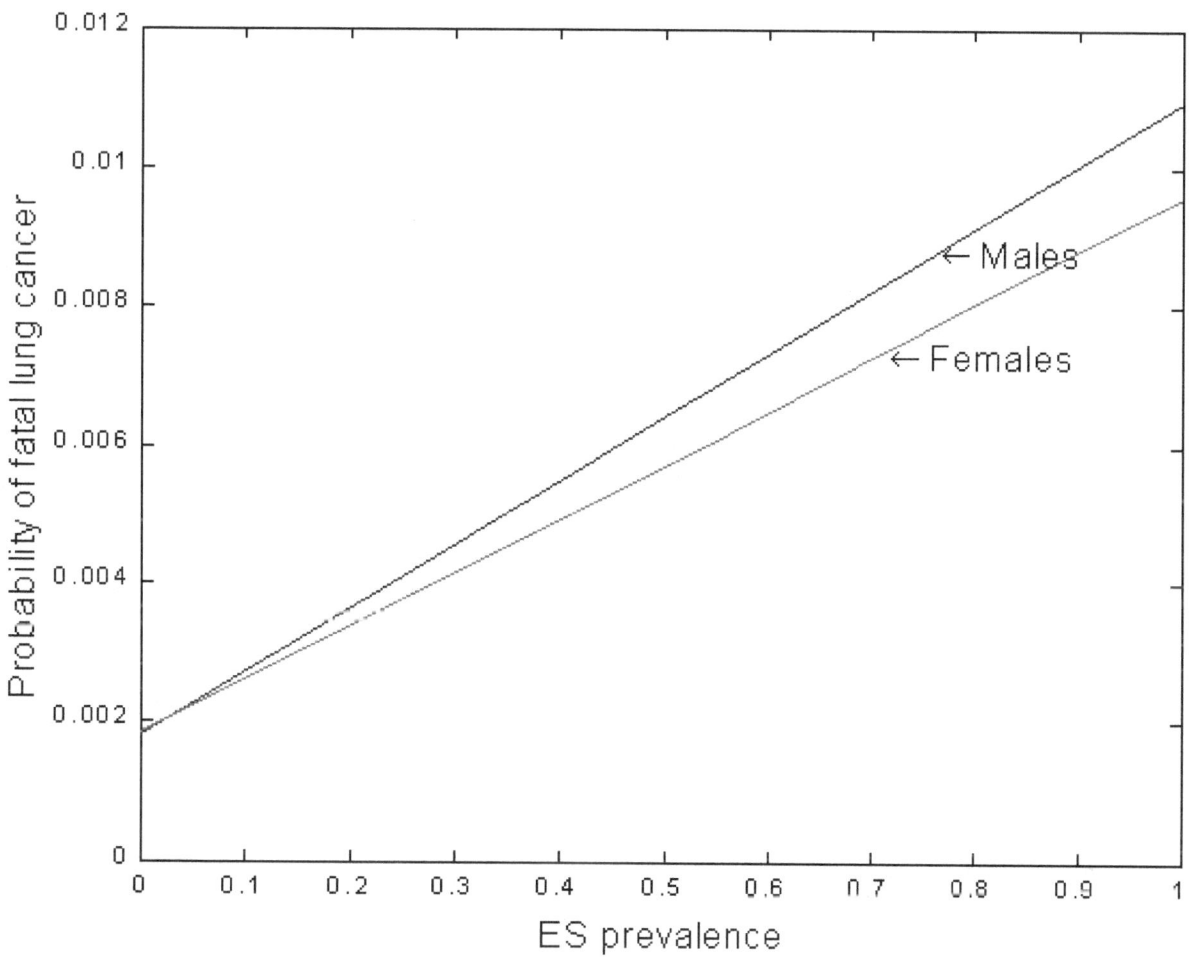

Figure 8: Probability of a premature lung cancer death from a lifelong exposure to radon at 1 pCi/L as a function of ES prevalence.

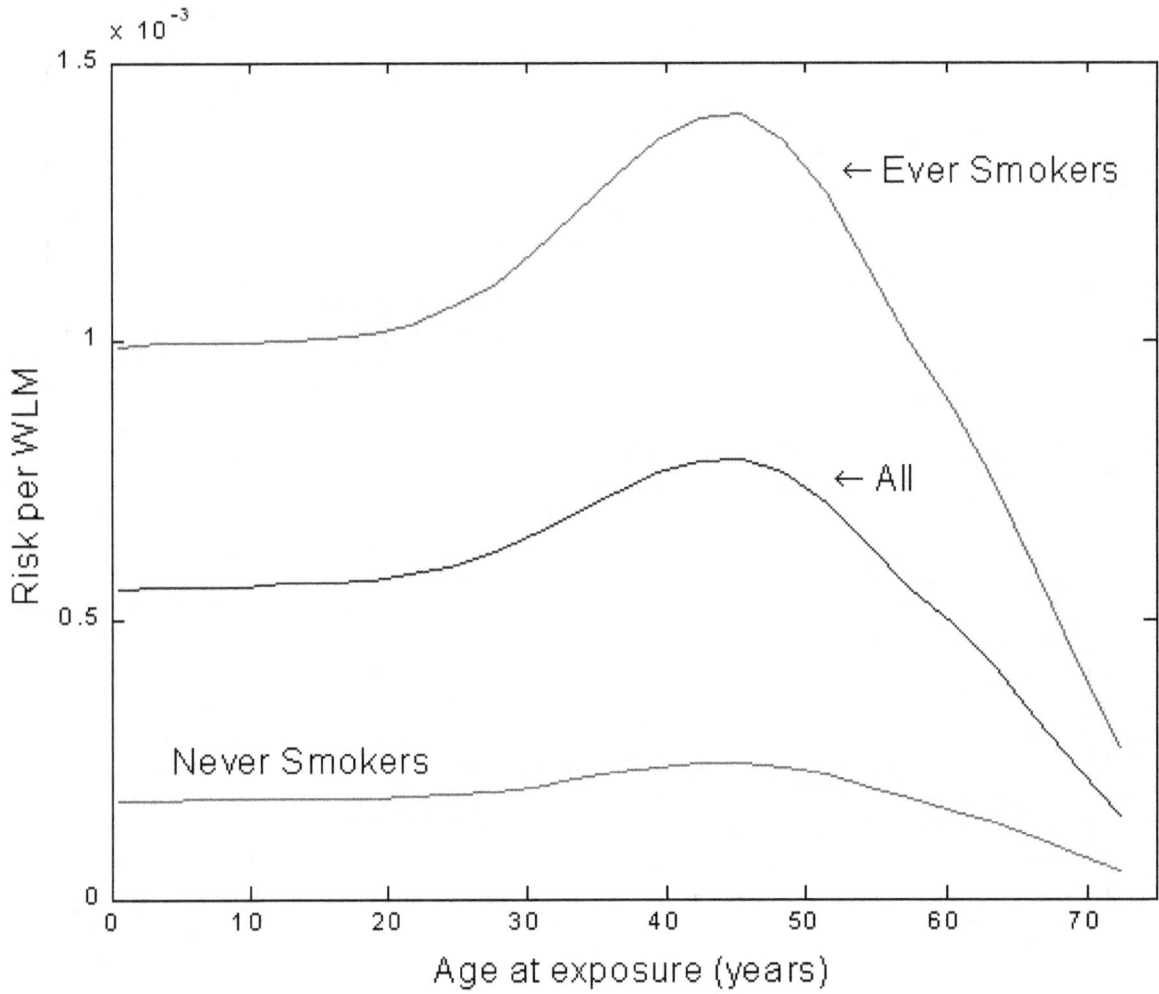

Figure 9: Risk per WLM as a function of age at exposure.

E. Age at Cancer Death and Years of Life Lost

Table 11 shows that, according to the concentration model, radon-induced lung cancer deaths tend to occur earlier than other lung cancer deaths. The estimated average age for radon-induced lung cancer deaths is about 65 y compared to 72 y for all lung cancer deaths. Years of life lost per death would then be greater for radon-induced lung cancers than other cancers, as Table 12 shows. For both males and females, the concentration model predicts an average of about 17 y of life lost per death when the cancer is radon-induced.

Age at lung cancer death and years of life lost per death depend on the shape of the ERR as a function of attained age (see Figure 3), but not the scaling. Figure 10 shows the probability density function for years of life lost for the three different relative risk models discussed in BEIR VI. Since the shapes of the concentration and duration ERR functions are so similar, the resulting density functions for *YLL* are also similar. Not surprisingly, the average *YLL* estimated using the duration model is also about 17 y. In both cases, the ERR function is relatively large for young ages (between 35 and 55 y), implying a greater likelihood that radon-induced cancers occur earlier. In contrast, the constant relative risk function predicts relatively few early cancers, and as a result the constant relative risk function would predict fewer *YLL* (about 12 y). Our "best" estimate of average *YLL* is 17 y, which is derived using either of the two BEIR VI preferred models. In contrast, the constant relative risk estimate of 12 y presents a reasonable lower bound for describing the uncertainties in this estimate.

Figure 11 indicates how *YLL* depends on age at exposure. For both males and females, *YLL* appears to be relatively constant for exposures up to about age 40 y, and then *YLL* decreases with age.

Table 11: Estimated average age at lung cancer death.

Gender	All Lung Cancer Deaths	Radon-Induced Deaths
Males	70.6 y	64.5 y
Females	73.1 y	66.1 y
Both	71.7 y	65.2 y

Table 12: Estimated years of life lost per lung cancer death.

Gender	All Deaths	Radon-Induced Deaths	
	Average	Average	Median
Male	13.2 y	16.1 y	14.9 y
Female	14.4 y	18.6 y	17.6 y
Both	13.7 y	17.2 y	16.4 y

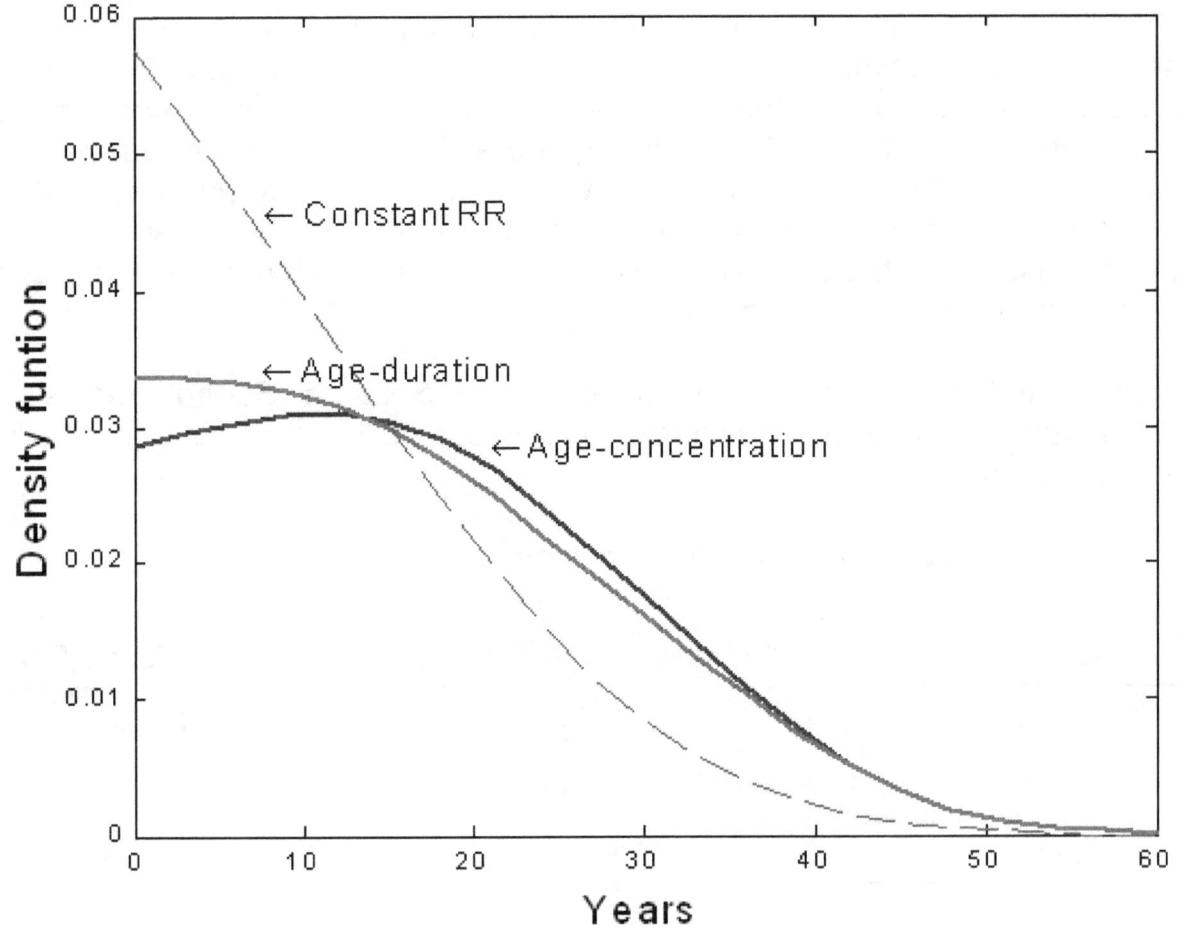

Figure 10: Density function for years of life lost from a radon-induced death

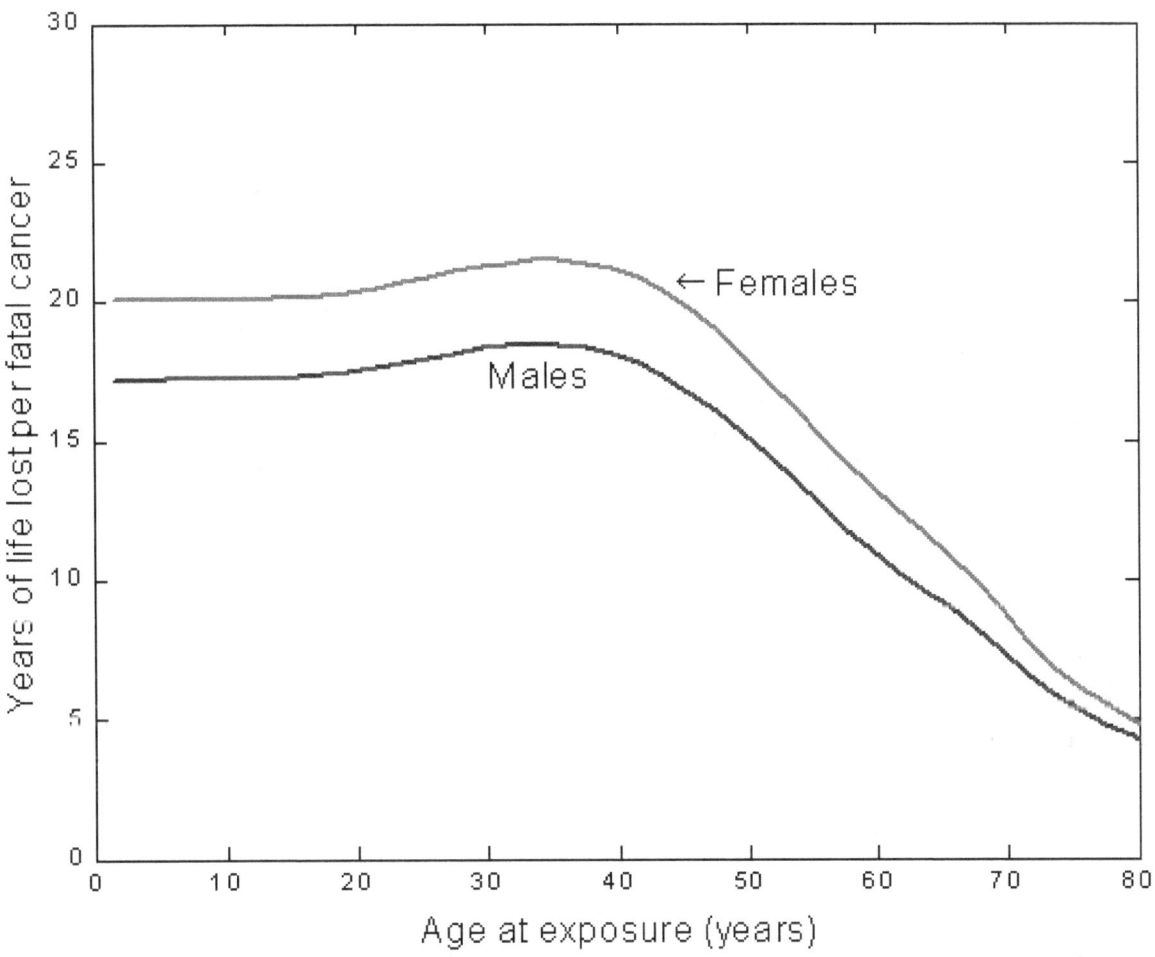

Figure 11: Years of life lost per fatal radon-induced cancer. Estimates based on scaled concentration model for exposures of one year duration as a function of age at exposure (midpoint of exposure interval).

F. Comparison with Previous Estimates

Our current estimate of risk per WLM (5.38×10^{-4}) is more than double the previous EPA estimate: 2.24×10^{-4} (EPA 1992). The corresponding proportion of lung cancer deaths estimated to have been radon-induced also increased, from 8.5% to 13.4%. Table 13 illustrates how changes in exposure parameters, the determination of baseline lung cancer rates, mortality data used, and the relative risk model, affected our estimates of risk per WLM and etiologic fraction.

1. Exposure parameters: The average radon daughter exposure rate is 0.181 WLM y^{-1}. This is based on BEIR VI determinations that: (1) on average, people spend 70% of their time indoors at home (occupancy factor, $\Omega = 0.70$) and (2) in homes, the average equilibrium fraction for radon daughters is $F = 0.4$. Taken together with the estimated average radon concentration of $C=1.25$ pCi/L in the U.S. (Marcinowski *et al.* 1994), the estimated average exposure rate is:

$$w = C [F \times 0.01 \text{ WL } (\text{pCi/L})^{-1}] [\Omega \times 51.6 \text{ WLM } / (\text{WL-y})^{-1}]$$

$$= (1.25 \text{ pCi/L}) [(0.7)(0.4) \text{ WL } (\text{pCi/L})^{-1}] [0.516 \text{ WLM } (\text{WL-y})^{-1}]$$

$$= 0.181 \text{ WLM/y}$$

This value is about 25% lower than EPA's previous estimate of 0.242 WLM/y based on $\Omega = 0.75$ and $F = 0.5$.

Changing the exposure from 0.242 WLM/y to 0.181 WLM/y has little effect on risk per WLM but decreases the *EF* almost proportionally. EPA's previous risk estimate was 2.24×10^{-4} per WLM with an etiologic fraction of about 8.5%. Based on the same 1992 assumptions, but using 0.181 WLM/y, the estimated risk per WLM would be essentially unchanged, but the estimated etiologic fraction would be 6.5%.

2. Baseline rates: Because EPA's radon risk estimates are determined using relative risk models, they depend directly on the baseline lung cancer death rates that are used. In 1992, we adjusted the observed lung cancer death rates downward to obtain baseline rates from all causes *other* than residential radon exposure. For our current risk estimates, we have not adjusted the baseline rates. A discussion of this issue is given in (Nelson et al. 2001).

Table 13 indicates that not adjusting the baseline rates increases both the risk per WLM and the etiologic fraction. With previously used mortality data and relative risk models, an exposure of 0.181 WLM/y and adjustment of the baseline rates, the estimates of risk would be 2.3 per 10,000 WLM with an etiologic fraction of 6.5%. With the same

44

inputs and relative risk model, but no adjustment of the baseline rate, the risk per WLM would be 2.5 per 10,000 WLM with an etiologic fraction of 7.0%.

3. Mortality data: The previous EPA estimates were based on 1980 mortality data. Updating the mortality data to 1990 increases the risk per WLM and causes a slight decrease in the calculated etiologic fraction. The risk per WLM increases because of the increases in baseline lung cancer rates. The etiologic fraction most likely decreased because stationary populations based on 1990 data contain a greater proportion of older people, and radon-related relative risks decrease with attained age.

4. Relative risk model: In 1992 EPA used the BEIR IV relative risk model:

$$ERR(a) = 0.0175 \ \gamma(a) \ (W_1 + \tfrac{1}{2} W_2).$$

The predicted age-specific excess relative risks from the BEIR IV model tend to be only about half as large as the ERR from the scaled concentration model. As a result, switching to the concentration relative risk model roughly doubles both the estimates of risk per WLM and etiologic fraction (see Table 13).

Table 13: Dependence of risk estimates on changes in methodology since 1992.

Exposure Rate (WLM /y)	Adjustment of Baseline Rates	Mortality Data	Relative Risk Model	Risk per WLM	Etiologic Fraction
0.242	Yes	1980	BEIR IV	2.2	8.5%
0.181	Yes	1980	BEIR IV	2.3	6.5%
0.181	No	1980	BEIR IV	2.5	7.0%
0.181	No	1990	BEIR IV	3.0	6.7%
0.181	No	1990	Scaled Concentration	5.4	13.4%

G. Effects Other than Fatal Lung Cancers

The estimates above refer only to fatal lung cancers. As cited in BEIR VI, lung cancer incidence in 1994 was estimated to be about 12% higher than lung cancer mortality (NAS 1999, DHHS 1995). Assuming that the etiologic fraction would be nearly the same for lung cancer incidence as mortality, the numerical estimates of risk per WLM and of radon-induced lung cancers would be about 12% higher than for lung cancer mortality. Thus, one might project about 23,600 (21,100×1.12) radon-induced lung cancer cases in 1995.

To a limited extent, inhaled radon gas is absorbed into the bloodstream and transferred to all parts of the body. Radioactive decay of this radon and its daughters result in a radiation dose—predominantly from alpha-particles—to all potential cancer sites. The cancer risk associated with this dose is very small compared to the lung cancer risk due to decay of radon decay products deposited in the bronchial epithelium. Using dosimetric models and risk factors recommended by the International Commission on Radiological Protection, James (1992) has estimated that the risk to other organs is about 2% of the lung cancer risk.

H. Current Smokers

The BEIR VI committee provided relative risk models and estimates of radon-induced lung cancer deaths for both ES and NS but did not provide guidance on how the risks may differ for current versus former smokers. Most likely, this is because the miner data provide relatively little information on how to distinguish between the former and current smoker risks. An approach to this issue is to assume that the radon-related relative risk given by the concentration model is the same for former and current smokers; that is, the relative risks for both former and current smokers are assumed to be about 0.9 times the relative risks for the U.S. population. Taking into account the respective baseline lung cancer death rates in the two groups, this made it possible to provide the rough estimates, shown in Table 14, of the number of radon-induced lung cancer deaths in 1995. The same relative risk assumptions were used to produce the risk per WLM estimate for current smokers described later in this section.

To derive the estimates in Table 14, we first partitioned the number of ES lung cancer deaths among former and current smokers, and then applied the ES etiologic fractions of 0.129 or 0.116 given in Table 8 for males and females, respectively. The partitioning can be accomplished as follows. First, age-specific lung cancer death rates can be obtained using 1990 vital statistics and assuming that the lung cancer death relative risks are 27.05 for male current smokers, 10.69 for male former smokers, 13.45 for female current smokers, and 4.47 for female former smokers (Malarcher et al. 2000). Second, using 1990 census data and prevalence data for current and ever smokers (DHHS 1997), we can estimate age-specific numbers of current and former smokers for both males and females. By then applying the age-specific lung cancer death rates to these numbers of former and current smokers, we have calculated that about 50% of male and 67% of female ES lung cancer deaths in 1990 were among current smokers. Assuming these percentages were similar for 1990 and 1995, about 45,300 of the 90,600 male and 37,300 of the 55,800 female ES lung cancer deaths in 1995 were among current smokers. For males, although the risks of lung cancer are about 2-3 times greater for current than former smokers, the surprisingly high number of lung cancer deaths among former smokers is due to a much higher former smoker than current smoker prevalence at ages at which cancer is most likely to occur.

Relative risks for current and former smokers are based on data from the Cancer Prevention Study II (CPS II). The CPS II is a large cohort study of about 1.2 million participants, who were recruited by volunteers from the American Cancer Society. The CPS II participants are not representative of the general U.S. population. However the Office of Smoking and Health (OSH) of the Centers for Disease Control has carefully analyzed the CPS II data to ensure the validity of risk estimates that result from the survey (Malarchar *et al.* 2000).

Table 14: Estimating radon-induced lung cancer deaths for current and former smokers.

Gender	Smoking Category	Lung Cancer Deaths		Fraction Due to Radon	Radon-Induced Lung Cancer Deaths
Male	Ever	90,600		0.129	11,700
	Current	50% of ES	45,300	0.129	5,850
	Former	50% of ES	45,300	0.129	5,850
Female	Ever	55,800		0.116	6,500
	Current	67% of ES	37,300	0.116	4,300
	Former	33% of ES	18,500	0.116	2,200

Since they have a higher baseline rate, it is likely that the the risk per WLM would be greater for current smokers than ever smokers. Using the same life-table methods described in Section VI.B, and relative risk values of 27.05 for males and 13.45 for females it is possible to calculate a crude estimate of risk per WLM for current smokers (presumed to start smoking at age 18 y and do not quit) equal to 15×10^{-4} (rounded to the nearest 5×10^{-4}). This estimate suggests about a 50% greater risk per WLM for lifelong smokers than ES.

It should be emphasized that the estimates given in this section may be especially sensitive to assumptions on smoking, including some that were not needed in BEIR VI (because the committee confined estimates to NS and ES). The next section offers a discussion on estimation problems related to smoking.

I. Dependence of Lung Cancer Death Rates on Smoking

The validity of the life-table calculations as estimates of risk for present and future radon exposures depend on several factors, including whether mortality rates — especially lung cancer death rates — remain reasonably stable. Possible changes in lung cancer rates must be considered because these rates are extremely sensitive to changes in smoking prevalence and habits. Smoking patterns have changed and will continue to evolve, and these changes will undoubtedly affect future risks related to radon exposure.

For example, in 1990, the proportion of Americans of ages 25 to 44 y who had ever smoked 100 cigarettes (ES) was about 50%, versus 60% for adults of ages 45 to 64 y (CDC 1994). To account for the complicating effects of changing smoking patterns on mortality rates, we must make separate life-table calculations for ES and NS. As in BEIR VI, our separate life-table calculations have only accounted for the differential mortality effects of smoking-related lung cancer, but it has been estimated that lung cancer accounts for only about 28% of current smoking-related deaths in the U.S. (Bergen and Caporaso, 1999). Other major health effects of smoking include an increased risk of circulatory disease, benign lung disease such as emphysema, and other cancers. To determine whether we need to account for the differential mortality effects from all smoking-related diseases, we have made preliminary risk per WLM calculations for ES and NS, using life tables (Rogers and Powell-Griner 1991) for heavy, light, former and never smokers that accounted for the increased smoking-related risks of death from all causes (not just lung cancer). The tables had been derived from three national surveys, the 1985 and 1987 National Health Interview Surveys, and the 1986 Mortality Followback Survey. The overall risk per WLM derived this way differed only slightly from the risk per WLM already described in Section D and shown in Table 10. We can then conclude that the risk per WLM is not very sensitive to differential mortality due to smoking-related causes other than lung cancer. For simplicity, we have decided not to consider this issue further.

More vexing problems are suggested by results from the Cancer Prevention Studies, which showed that relative risks of lung cancer death associated with smoking change over time (DHHS 1997). For the period 1959-65, the estimated relative risk for current smokers was 11.9 for males and 2.7 for females. This means that among males, the lung cancer rate for current smokers was 11.9 times as great as that for never smokers, and for females the ratio was 2.7. For 1982-88, estimated relative risks were 27.1 for males and 13.5 for females. Factors that influence these relative risks include cigarette composition, number of cigarettes smoked, and smoking duration. Similarly, changes among former smokers, including trends in time since cessation, would affect both former and ever smoker relative risks.

Even if smoking patterns were stable, determining the relationship between smoking and lung cancer rates would still be complicated. Results from national studies, given in Table 15, indicate that the relative risks may depend on age at expression (Malarcher *et al.* 2000). From CPS II, the relative risks for current smokers tend to decrease with age at expression. A consistent trend in age-specific relative risks is not as evident from estimates derived by using data from both the National Mortality Feedback Survey (NMSF) and the National Health Interview Survey (NHIS). The NMSF collected data on a representative sample of decedents of ages 25 years or older; the NHIS is a nationally representative household survey. Results based on data from NMFS and NHIS may not be as reliable as from the CPS II, as indicated by the wider confidence intervals, and since the NMSF had to rely on proxy respondents for information on smoking. For a more comprehensive discussion, see Malarcher *et al.* (2000).

Unfortunately, at this time, one can not reliably predict how smoking-related relative risks may change over time, or quantify the dependence on age at expression. We have therefore decided to follow the recommendations of BEIR VI, which assumed that the relative risk of lung cancer death for ES is 14 for males and 12 for females, independent of age. Table 16 suggests that our estimates of risk per WLM for the general population and ES may be somewhat insensitive to assumptions about the relative risk of fatal lung cancers for ES. For relative risks ranging from 9.33 to 28 for males and 8 to 24 for females, the risk per WLM would be within 6% of the nominal estimates for either ES or the general population. In contrast, the estimated risk per WLM for NS would range from 0.9×10^{-4} to 2.4×10^{-4}.

Table 15: Age-specific relative risks[a] and age-adjusted relative risks of fatal lung cancers for current and former smokers[b] versus never smokers for whites (from Malarcher et al. 2000).

Sex	Age (y)	NMFS/NHIS[c]		CPS II[d]	
		Current	Former	Current	Former
Male	35-59	82.05 (19.8, 339)	27.96 (6.38, 122)	27.21 (16.5, 44.8)	11.09 (6.65, 18.5)
	60-69	10.73 (4.26, 26.7)	3.52 (1.37, 7.03)	30.71 (21.4, 44.0)	11.25 (7.82, 16.2)
	70-79	8.78 (3.65, 21.1)	3.86 (1.62, 9.19)	27.23 (19.6, 37.9)	9.43 (6.77, 13.1)
	≥ 80	19.25 (4.22, 87.8)	8.92 (2.04, 39.0)	13.40 (8.18, 21.9)	6.55 (4.15, 10.3)
	Age adjusted	40.65 (15.7, 105)	14.33 (5.33, 38.6)	27.05 (19.3, 37.9)	10.69 (7.57, 15.08)
Female	35-59	32.13 (7.56, 137)	12.77 (2.71, 60.1)	14.77 (10.8, 20.2)	4.53 (3.16, 6.50)
	60-69	11.22 (4.64, 27.1)	5.65 (2.13, 15.0)	14.70 (11.7, 18.5)	5.05 (3.88, 6.55)
	70-79	21.70 (9.50, 49.5)	7.50 (3.09, 18.2)	11.28 (8.88, 14.3)	4.50 (3.44, 5.90)
	≥ 80	27.19 (11.1, 66.4)	3.68 (1.28, 10.5)	7.31 (4.76, 11.2)	2.95 (1.81, 4.83)
	Age adjusted	24.39 (10.2, 58.4)	9.15 (3.59, 23.3)	13.45 (11.1, 16.3)	4.47 (3.58, 5.59)

[a] Central estimates of relative risks with 95% confidence intervals given in parentheses.

[b] Malarcher et al. define current smokers as persons who reported they smoked now; former smokers reported they had ever smoked but did not smoke now.

[c] NMFS, National Mortality Feedback Survey; NHIS, National Health Interview Survey

[d] CPS II, Cancer Prevention Survey II

Table 16: Sensitivity of risk per WLM estimates to assumptions about the relative risk of fatal lung cancers for ES compared to NS.

Ratio of ES divided by NS fatal lung cancer rates		Risk per WLM (10^{-4}) by smoking category		
Male	Female	ES	NS	All
9.33	8	9.39	2.41	5.64
14	12	9.68	1.67	5.38
21	18	9.89	1.15	5.19
28	24	10.0	0.88	5.10

J. Summary

We described three lifetime radon-related risk estimates for stationary populations based on 1990 U.S. mortality rates: the risk per WLM (5.38×10^{-4}), *EF* (about 0.134 in 1995), and *YLL* (17.2 y). Our estimates of risk per WLM are much larger for ES than NS, but *EF* is about twice as large for NS than for ES. These estimates are based upon a scaled version of the BEIR VI concentration model, assumptions that exposure to radon are constant, and assumptions about ES prevalence and smoking related health effects. We have discussed many of the ways these estimates depend on these assumptions and have shown, for example, that estimates of risk per WLM may be more sensitive to assumptions about smoking prevalence than estimates of *EF*. We have also presented risk estimates for specific ages at exposure with the caveat that these estimates are subject to considerable uncertainties. These include what might be termed "modeling uncertainties". Almost all risk estimates, including those described in this document, are dependent on the modeling framework used for the data analysis. The BEIR VI committee used relative risk models to analyze the miner data. Alternative models for the analysis of the miner data will be one of the topics of the next chapter.

VII. UNCERTAINTIES

A. Background

The BEIR VI committee identified 13 sources of uncertainty in its estimates of risks from indoor radon. These were divided into two categories: (1) uncertainties in the parameter estimates for the exposure-response model derived from the miner data and (2) uncertainties in specifying the form of the model and in its application to the general U.S. population. A quantitative uncertainty analysis was performed, but it was limited to those factors that could be addressed without relying heavily on the subjective judgment of experts. Thus, the quantitative analysis considered only the statistical variability in the miner data and the comparative dosimetry between mine and residential exposures. Employing a "random-effects model," the committee incorporated variation among cohorts as well as sampling variation within cohorts.

The BEIR VI committee provided quantitative uncertainty estimates for the AR and for the number of radon-induced lung cancer deaths, based on each of the models derived from the miner data. For the concentration model, the 95% confidence interval around the central estimate of AR (14%) ranged from about 10 to 27%. For the duration model, the central estimate was about 10% and the uncertainty interval ranged from about 8 to 20%. The committee also considered a simple constant relative risk model (CRR), which was based on an analysis of only those miners receiving an estimated exposure of less than 50 WLM. Although the CRR model was deemed to have less credibility than the duration or concentration model for calculating central estimates of risk and lung cancer deaths, the committee's preferred uncertainty estimates were obtained from the CRR model. The CRR analysis led to an uncertainty range of 2-21%, with a central estimate of about 12%.

The most striking difference among the projections is in the lower bound estimate for the CRR model, which is much lower than for the other two models. This difference primarily results from larger sampling errors inherent to the CRR estimate. Limiting the study population to miners with low exposures sharply reduces both the number and the attributable risk of radon-induced lung cancers, causing a large increase in the relative standard error. In stating its preference for the CRR model estimate of uncertainty, the BEIR VI committee notes that the low radon exposure conditions included in the CRR analysis are more comparable to those in homes.

We believe that the BEIR VI CRR model-based uncertainty analysis should be interpreted with caution since it excludes the useful information from miners with exposures greater than 50 WLM. In particular, the CRR analysis excludes available information about relationships between the dose response and modifying factors such as time-since-exposure or attained age. The much wider uncertainty intervals derived using the CRR approach appear to be a consequence of an arbitrary cutoff leading to a substantial increase in sampling error. There seems to be insufficient justification to reduce the lower bound estimate well below what was derived from analyses of the entire data set. Although the BEIR VI CRR-model uncertainty bounds provide a useful indication of how our estimates depend on data from miners with exposures ≤50 WLM, we believe that, at this point, the use of the scaled concentration model for deriving uncertainties is more consistent with our overall approach.

It should be noted that the uncertainty ranges derived in BEIR VI, based on application of the preferred models to the entire data set, do not fully reflect the degree of uncertainty because important sources of uncertainty were not factored into the quantitative uncertainty analysis. Sections B and C, below, discuss sources of uncertainty not treated quantitatively in BEIR VI.

As in BEIR VI, we have generally limited our quantitative uncertainty analysis to factors that can be addressed without relying heavily on subjective expert judgement. The quantitative uncertainty analysis relies on a Monte Carlo simulation that accounts for uncertainties in residential exposures (see Section D), uncertainties in parameters values in the BEIR VI concentration model, the K-factor, and a scaling factor (included because of differences between results from the BEIR VI duration and concentration model). Major differences between our Monte Carlo simulation and the simulation used in BEIR VI is in the way we treat uncertainties in residential exposures, the K-factor, and the fact that the BEIR VI simulations do not account for the additional scaling factor. Neither our simulation (see Section E) nor the BEIR VI simulation account for uncertainties in extrapolating to low exposure rates — see Section F for a discussion. Finally a very simple sensitivity analysis is given in Section G to indicate how our estimates may depend on assumptions about risks from exposures to ES, NS, and children, and how relative risks depend on time-since-exposure.

B. Uncertainties in the Miner Data

1. Errors in exposure estimates: There are two such major issues with respect to the miner data itself. First, the data on miner exposure is deficient in many ways, which may bias the estimate of the relative risk coefficient to varying degrees in the individual cohort studies. Moreover, as stated in BEIR VI (NAS 1999, p. 161): "For most of the cohorts, exposure measurement errors are likely to be greatest in the earliest periods of operation, when exposures were largest and fewer measurements were made. For this reason, measurement errors not only affect the estimates of the overall risk coefficient, but may also bias estimates of parameters that describe the relationship of risk with other variables such as exposure rate, time-since-exposure, and age at risk." Since the magnitude of possible errors in exposure estimates are often extremely difficult or impossible to quantify, it is very hard to estimate the magnitude of the uncertainty in risk estimates introduced by this source. The reasonable concordance among the various miner studies is somewhat reassuring on this point; in particular, removal of any one study from the analysis has little effect on the overall risk estimate. Nevertheless, differences in the ERR/WLM estimated from the various miner studies are larger than what could be expected from sampling errors alone, and it is likely that the exposure errors do contribute substantially to these differences.

2. Confounding by other exposures: Second is the issue of possible confounding with other mine exposures. Some miners were exposed to arsenic, silica, and diesel exhaust, all of which may affect lung cancer risk. The BEIR VI report concludes that diesel exhaust appears to be a weak carcinogen and is "probably not a strong modifier of the risk of radon progeny." The data on silica are somewhat conflicting, and silica's role has not been directly assessed, but the scant epidemiological evidence does not show silicosis to be a strong modifier of radon risk (Samet *et al.* 1994, NAS 1999). Two of the miner cohorts, China and Ontario, had quantitative data on arsenic exposure; in addition, Ontario, Colorado, New Mexico, and France had data indicating whether miners had previous mining experience. Adjusting for arsenic exposure in the Chinese cohort sharply reduced the estimate of ERR/WLM from 0.61% to 0.16%. Otherwise, adjustment for arsenic or previous mining made little difference in the estimated risk coefficient.

In evaluating the possible effect of other exposures on the estimated ERR/WLM, it is important to consider whether these exposures are correlated with radon progeny exposures and whether they act synergistically with radon in causing lung cancer. So, for example, in the case of the Chinese tin miners, arsenic and radon exposures were highly correlated; moreover, when the data were adjusted for arsenic exposure, the ERR/WLM was similar across arsenic exposure categories, suggestive of a multiplicative interaction between the two carcinogenic agents. Failure to adjust for arsenic exposure, in this case, would have led to an overestimate of the ERR/WLM. Were it to be determined that the interaction is actually submultiplicative, the estimated ERR/WLM would have to be increased from 0.16% to a value between 0.16% and 0.61%.

In contrast, if there were a multiplicative interaction but no correlation between exposures, no bias in the ERR/WLM would result. For example, little correlation between radon and cigarette smoking is expected; furthermore, the two agents are strongly synergistic in causing lung cancer. Therefore, confounding by smoking should be small. Nevertheless, uncertainties in miner smoking may produce considerable uncertainty in radon risk estimates. This uncertainty results not from confounding, but from the lack of detailed smoking information.

On the other hand, if there were no correlation and no synergism (*i.e.*, the effect of the two exposures is additive), then the other exposure will simply produce a uniform increase in lung cancer rates across radon exposure categories, resulting in an underestimate of the ERR/WLM.

3. Smoking by miners: Five miner cohorts had useful information on smoking: China, Colorado, Newfoundland, Malmberget, and New Mexico. From this restricted data set it was determined that the interaction between radon and smoking was probably submultiplicative, although a multiplicative interaction could not be excluded. Overall, a best fit to the data indicated that NS had about 2.1 times the ERR/WLM as ES. There are wide uncertainty bounds on the risk estimate for NS, and therefore on the ratio of the risk coefficients for the two groups. Some perspective on the uncertainty can be gained by examining the effect of omitting any one of the cohorts from the analysis (NAS 1999). Omitting the Chinese cohort data had the largest percentage effect, reducing the ratio from 2.1 to 1.2. The largest increase was found when the New Mexico cohort was omitted; in that case the estimated ratio increased to almost 3. While the BEIR VI committee did not provide any quantitative uncertainty estimates by smoking category, it would appear that there is roughly an extra factor of two uncertainty in the risk for NS compared to that for ES (or for the general population).

There is very little information on risks to miners who had ceased smoking, and BEIR VI does not explicitly quantify the risk to former smokers. Former smokers are subsumed in the ES category. Using the BEIR VI model and baseline lung cancer rates for former smokers, risk estimates for this group could be derived. Estimates for former smokers would be more uncertain than those for ES, as a group; moreover, the relative risk for individual former smokers may vary greatly with detailed smoking and radon exposure histories.

C. Uncertainties in Extrapolating to Residential Exposures

1. _K_-factor: In performing its quantitative uncertainty analysis, the BEIR VI committee did consider variability and uncertainty in the _K_-factor, concomitant with the uncertainties in modeling the miner data. The variability in _K_ was characterized as a lognormal distribution with a gm=1.0 and a gsd=1.5. The gsd itself was assigned an uncertainty distribution that was loguniform over the range from 1.2 to 2.2, _but no uncertainty in the gm was assumed._ When the variability in _K_ was incorporated into the quantitative uncertainty analysis, the uncertainty distributions were shifted upwards. Basically, the reason for the shift is that the mean value of the distribution for _K_ is higher than the gm, which was used as the nominal estimate. Factoring in the uncertainty in the gsd had little effect on the respective lower bound estimates but did lead to significant increases in upper bound and median estimates (NAS 1999: Table A-10).

The BEIR VI committee treated the uncertainty associated with the median estimate (gm) for _K_ as negligible. While the variability in _K_ is larger than the uncertainty in its median value, this seems unreasonable in view of the sensitivity of the K-factor to how the respiratory tract is modeled and our imperfect knowledge of the parameters affecting estimates of _K_, such as aerosol size distributions, ultrafine fractions, breathing rates, nasal deposition of the ultrafine activity, and relative radiosensitivity of lung regions. In particular, only limited information is available for estimating aerosol conditions in mines without diesel engines despite the fact that many miners in the epidemiologic studies worked in mines without such equipment (Cavallo 2002). Correspondingly, the aerosol size distributions in homes were based on measurements in just 6 homes (NAS 1999).

Cavallo (2000) argued that it is likely that the _K_-factor is much less than 1, because in mines "more particles are found at larger or smaller diameters where the deposition and dose per unit of progeny concentration is substantially higher." James et al. (2003)

calculated a value of K of about 1 based on the exposure assumptions described in BEIR VI. Other results (Porstendörfer and Reineking 1999) indicate that the K-factor could be somewhat larger than 1. Using a lung dose model "with a structure that is related to the new ICRP respiratory tract model", their estimate of dose per unit exposure is about 1.1 times as large for homes with "normal" aerosol conditions as compared to mines without working activities. For homes without smoking, their calculations indicate a K-factor as large as 1.5, suggesting a sensitivity to aerosol conditions in the home. To be consistent with the range of K-factors that these results and arguments suggest, we have subjectively assigned a normal distribution with $\mu = 1.0$ and $\sigma = 0.25$ for the median value of the K-factor.

2. Dependence of risks on gender: The miner cohorts used to develop the BEIR VI risk models consisted of males only. The BEIR VI committee assumed that the risk models derived for male ES and NS apply equally to females. This seems reasonable. The estimated K-factor is almost the same for females and males (NAS 1999). Baseline lung cancer rates are substantially lower for females than for males, but this reflects differences in past smoking patterns; over time, lung cancer rates in females are approaching those in males. Indeed there is now evidence suggesting that females are more susceptible to tobacco carcinogens (Zang and Wynder 1996). It is unclear how such differences in susceptibility would be reflected in radon risk; it would depend on whether the degree of synergism between smoking and radon was appreciably different for the two sexes. Given the lack of information on this point, radon risks for females must be regarded as more uncertain.

3. Dependence of risks on age at exposure: Essentially all the data on childhood exposures to radon were obtained from the Chinese tin miner cohort, and even those data are relatively sparse. Consequently, the uncertainty in risks associated with childhood exposures must be regarded as substantially higher than for adult exposures. As shown in Table 17, the ERR/WLM observed for the Chinese miners who began mining as children is generally about a factor of two higher than for others in the cohort, even after adjustment in the background risk for various other factors (Xuan et al. 1993). Since any enhanced effect of childhood exposures among these miners would have been diluted by the effect of additional exposures received as adults, these results suggest that the relative risk coefficient associated with childhood exposures could be several times higher than for adult exposures. These findings must be interpreted with caution, however. Xuan et al., as well as others who have examined these results (Lubin et al. 1994, NAS 1999), concluded that the pattern of risk did not vary consistently with age at first radon exposure.

Table 17: Effects of age at first radon exposure on ERR per WLM for several analyses of a Chinese tin miner cohort (from Xuan et al. 1993).

ERR adjustment[a]	Age at first radon exposure (y)					
	<10	10-14	15-19	20-24	25-29	≥30
None	1.0[b]	1.1	1.1	0.2	0.4	0.7
Attained age	1.0	1.1	0.9	0.3	0.5	0.4
Time-since-exposure	1.0	1.3	1.1	0.3	0.6	0.5
Radon rate	1.0	1.2	1.2	0.4	0.5	0.7

[a] Each analysis adjusted ERR by either attained age, time-since-exposure, radon rate, or none of the above.
[b] Baseline risk fixed at 1.0
[c] Background risk adjusted for age and arsenic exposure.

There are also the unique features of the Chinese miner cohort that may make the results on these miners less applicable to the residential exposures of interest than those obtained from the other miner cohorts. In particular, lung cancer rates in the cohort are extremely high (38% of all deaths), probably due in large part to arsenic exposure (Xuan et al. 1993). Since arsenic and radon exposures are strongly correlated, the potential for confounding by arsenic is great. Moreover, these miners were unique in their smoking habits, using water pipes in addition to cigarettes. The nature of the interactions between the smoking, arsenic, and radon in causing lung cancer is problematic. It is also noteworthy that, after control for arsenic exposure, the estimates of ERR/WLM are substantially lower for this cohort than for the other miner cohorts. Thus, the estimated risk coefficients for the Chinese miners who began working as children are still lower than the estimated risk coefficients derived from the other miner studies.

Epidemiologic follow-up on the atomic bomb survivors fails to show any clear evidence of an enhanced risk of radiation-induced lung cancer associated with childhood exposures as is seen with some other cancer sites (Thompson et al. 1994). Moreover, both the miner data and the atomic bomb survivor data on lung cancer exhibit a falloff in the ERR with time after exposure, a falloff that is likely to continue beyond the period of epidemiologic follow-up. Although the radiation exposure was predominantly low-LET (γ-rays) in the case of the atomic bomb survivors rather than high-LET (α-particles), these considerations also tend to argue against highly elevated lifetime lung cancer risks from childhood radon exposures.

In conclusion, the uncertainties in risk estimates for childhood radon exposures are larger than for the general population but are difficult to quantify. Information on these risks could, in principle, be gained through residential studies, but carrying out such studies would be logistically difficult—and might introduce new uncertainties— because of the long time elapsing between exposure and onset of the disease.

4. Smoking patterns in the U.S. population: As shown in Figure 7, estimates of *EF* (*i.e.,* the fraction of lung cancers attributable to radon) are fairly insensitive to the baseline lung cancer rates or smoking patterns in the population. In contrast, projections of risk per unit exposure or of the number of radon-induced lung cancer deaths are strongly dependent on population lung cancer rates and smoking patterns. As a consequence, there is additional uncertainty in projecting population risks due to current and future radon exposures because of uncertainties with respect to future trends in lung cancer rates and smoking.

Another source of uncertainty relates to how the synergism between radon and cigarette smoke depends on the temporal pattern of the exposures. Animal data suggest that the exposures act synergistically only when the radon exposure precedes the cigarette smoke (Chemaud *et al.* 1981, Cross 1994). This would suggest that the risks of childhood radon exposures would be enhanced by adolescent or adult smoking. On the other hand, radon exposure to former smokers (subsequent to smoking cessation) may pose only about the same risk as exposures to NS of the same age. At this time, however, the risks from radon exposures received prior to starting, or subsequent to quitting, smoking remain highly uncertain.

D. Uncertainty in the Estimate of Average Residential Exposure

As noted in Section VI.F, the average annual residential exposure is estimated to be:

$$w = C\,[F \times 0.01\ \text{WL/(pCi/L)}]\,[\Omega \times 51.6\ \text{WLM (WL-y)}^{-1}]$$

where: *C* is the average radon concentration in homes, *F* is the average equilibrium fraction, and Ω is the occupancy factor (average time spent indoors at home). The nominal estimates adopted for *C*, *F*, and Ω are 1.25 pCi/L, 0.4, and 0.7, respectively, which imply a nominal average exposure rate of *w* =0.181 WLM/y. To evaluate the uncertainty in *w*, we must consider the uncertainty in each of the three parameters, *C*, *F*, and Ω .

1. Uncertainty in the average radon concentration (C): EPA's National Residential Radon Survey (NRRS) determined the average radon concentration to which people are exposed in their homes is 1.25 pCi/L, with a standard error in measurement of 0.06 (Marcinowski *et al.* 1994). However, this estimate was based on a simple arithmetic average of the concentration levels on floors "frequently occupied." This takes no account of: the fraction of time spent on each floor; the variability of radon level on a particular floor and how this is correlated with people's location in the house; nor the time spent on floors not classified as a frequently occupied areas (particularly basements). Floor occupancy data were collected in NRRS, but they primarily reflect summertime activity patterns, which are likely to be atypical; consequently, they were not used to adjust the estimate of average indoor radon. If they had, the estimate would have been reduced by about 7%. Based on these considerations, a normal distribution is assigned to C with μ =1.2 and σ =0.08.

2. Uncertainty in equilibrium fraction (F): The BEIR VI committee recommended a value of 0.4 for F, based on a detailed study of six homes (Hopke *et al.* 1995). Within these homes the equilibrium fractions were highly variable with time and dependent on the presence of a smoker. While this study was sensitive to temporal variations in F, it is unclear to what extent this small sample of homes is typical of U.S. residences. Based on measurements in 21 homes in New York or New Jersey, George and Breslin (1980) found that F was, on average, about 50% in basements and about 60% or higher on other floors; in contrast, an average equilibrium fraction of 33% was determined from measurements of 20 houses in Butte, Montana (Israeli 1985). A larger survey of livable areas in 200 houses conducted by the state of New Jersey yielded an average equilibrium factor of 45% (NJDEP 1989). To characterize the uncertainty in F, we have subjectively assigned a lognormal probability distribution, with gm=0.40 and gsd=1.15, corresponding to a 90% CI of 0.32-0.50.

3. Uncertainty in the average occupancy factor (Ω): A large survey of human activity patterns has recently been conducted by the EPA (Tsang and Kleipeis 1996). Results from this National Human Activity Pattern Survey (NHAPS) are summarized in EPA's 1997 *Exposure Factors Handbook*. The NHAPS data are compilations of 24-hour diary information from a sample selected using a random digit dial method. Results from the 9,386 respondents were weighted to obtain results representative of the U.S. population and of specific demographic factors, seasons, etc. It was found that, on average, Americans spend 67% of their time indoors in residences. The sampling error is estimated to be only about 0.3%. The response rate to the survey was 63%; of the remaining 37%, roughly two-thirds were contacted but refused to participate, and the remainder could not be contacted. The incomplete response may result in some error. In particular, the survey may have missed people who were away from home on vacation, etc. As a result, the estimate of Ω is likely to be biased high. Other errors may result from recall bias and imperfections in the sampling methodology. Taking into account these problems, we have assigned a normal probability distribution to Ω, with a mean of 0.65 and a standard deviation of 0.03.

E. Monte Carlo Simulation

We describe here a Monte Carlo simulation for quantifying uncertainties for estimates of risk per WLM, *EF, YLL,* and number of (radon-induced) fatal lung cancer deaths. The simulation is similar to those used by the BEIR VI committee, in that it is limited to factors that can be addressed without relying heavily on subjective expert judgement. The simulation accounts for uncertainties in factors for determining the average residential radon exposure, the *K*-factor, and uncertainties in the fitted parameter values in the BEIR VI concentration model for calculating age-specific ERR's.

Distributions for these parameters are given in Tables 18 and 19. Parameters for determining (miner-equivalent) doses, average radon concentration (pCi/L), occupancy factor, equilibrium factors, and the *K*-factor, are all assigned normal or lognormal distributions as detailed in the previous section. The BEIR VI concentration model for calculating the corresponding ERR's can be written as:

$$ERR\,(a) = \lambda\,\beta\,(w_{5\text{-}14} + \theta_{15\text{-}24}\,w_{15\text{-}24} + \theta_{25+}\,w_{25+})\phi(a) \qquad (6)$$

Here, we have explicitly included the scaling factor, λ, which was assigned the nominal value 0.825 to obtain risk estimates between the concentration and duration model estimates. In our simulation, λ has a lognormal distribution with gm = 0.825 and a gsd = 1.31, the gsd of the duration and concentration model *RWLM* estimates. This distribution for λ reflects the dependence of the modeled ERRs on the way either exposure duration or concentration are categorized. For the simulation, we assumed as in BEIR VI that the attained age function, $\phi(a)$, is constant within age intervals <55 y, 55 through 64 y, 65 through 74 y, and ≥75 y. As in the BEIR VI report, we assigned lognormal or normal probability distributions to the parameters β, $\theta_{15\text{-}24}$, θ_{25+}, $\phi_{55\text{-}64}$, $\phi_{65\text{-}74}$, and ϕ_{75+}, with median values equal to the fitted BEIR VI concentration model estimates and the covariances shown in Table 18.

The simulation of lifetime risk estimates began by repeatedly generating n = 10,000 sets of the exposure factors, *K*-factor, and relative risk model parameters. For each set of factors/parameters, we then calculated the average residential exposures and corresponding age-specific ERRs. The lifetable methods described in Section VI.B.4 were then used to calculate the risk per WLM, *EF*, and *YLL*. For all our simulations, we used 1989-91 mortality data and used age-specific ES prevalence data. As in the BEIR VI report, we multiplied the age-specific relative risks (for radon-induced deaths) by a factor of 2.0 for NS and 0.9 for ES. The relative risk for all lung cancer deaths between ES and NS was set to 14 (males) and 12 (females).

Results from the Monte Carlo simulation are summarized at the bottom of Table 19. The risk per WLM is about 2×10^{-4} to 12×10^{-4}, and the *YLL* is about 15 y to 20 y. The average annual residential exposure is between 0.12 WLM and 0.21 WLM, resulting in an *EF* greater than 0.05, corresponding to more than 8,000 radon induced fatal lung cancers. It seems highly unlikely that the *EF* or the number of radon-induced fatal lung cancers are as large as the calculated upper bound limits (0.30 and 45,000). Nominal estimates for the risk per WLM, *EF*, the number of premature lung cancer deaths, and *YLL* are all very close to the respective median values. To separate out the effect of uncertainties in exposure factors, the simulation was repeated with radon concentration, occupancy factor, and equilibrium factor set to the nominal values: 1.25 pCi/L, 0.7, and 0.4. Resulting uncertainty intervals (see Table 20) are 0.06 to 0.3 (*EF*) and 9,000 to 50,000 (number of premature lung cancer deaths). It appears that uncertainties in exposure factors are minor compared to uncertainty in factors that determine relative risk. For *EF*, the ratios of the endpoints of the uncertainty intervals are about 5, regardless of whether the exposure factors are simulated or held constant.

Results of the simulation are sensitive to the choice of distribution for λ, for which minimal information was available. The simulation does not account for uncertainties associated with errors in miner exposure estimates, confounding due to exposures other than radon in the mines, health effects related to ever-changing smoking habits, risks associated with childhood radon exposures, or model mis-specification. In our case, no model mis-specification would mean that the ratios of residential radon induced lung cancer mortality rates to background rates depend only on the exposures in WLM within intervals determined by the BEIR VI age-concentration categorizations for attained age and time-since-exposure. A discussion of alternative risk models that are biologically motivated is given in the next section. This is followed by a section that describes the sensitivity of our risk estimates to parameters that characterize and differentiate risks for subgroups such as ES, NS, and children.

Table 18: Parameters for uncertainty distributions for risk factors in the concentration model (NAS 1999)

I. Estimated Parameter Values					
Parameter Value	$\log(\beta)$	θ_{15-24}	θ_{25+}	$\log(\phi_{55-64})$	$\log(\phi_{65-74})$	$\log(\phi_{75+})$
	-2.76	0.77	0.51	-0.56	-1.23	-2.38

II. Covariance Matrix						
	$\log(\beta)$	θ_{15-24}	θ_{25+}	$\log(\phi_{55-64})$	$\log(\phi_{65-74})$	$\log(\phi_{75+})$
$\log(\beta)$	9.47					
θ_{15-24}	-0.36	0.77				
θ_{25+}	-0.04	0.24	0.42			
$\log(\phi_{55-64})$	-2.87	-0.10	-0.15	5.71		
$\log(\phi_{65-74})$	-3.18	-0.17	-0.33	2.85	10.87	
$\log(\phi_{75+})$	-3.44	-0.19	-0.54	2.90	3.20	87.65

Table 19: Monte Carlo simulation of Risk per WLM, EF, YLL, average residential exposure, and number of radon-induced fatal cancers.

I. Parameter Assumptions		
Exposure factors	Radon concentration (pC/L)	Normal (μ = 1.2, σ = 0.08)
	Occupancy factor	Normal (μ = 0.65, σ = 0.03)
	Equilibrium factor	Lognormal (gm = 0.4, gsd = 1.15)
K-factor		Normal (μ = 1.0, σ = 0.25)
Proportion of youth (< 18 y) that will smoke		0.37 (males); 0.36 (females)
Exposure response parameter ratios		0.9 (ES vs. All); 2.0 (NS vs. All)
Relative risks of lung cancer death from smoking; (ES vs. NS)		14.0 (Males), 12.0 (Females)
Age-concentration model parameters		See Table 18
Relative risk model scaling parameter		$\lambda \sim$ LN (gm = 0.825, gsd = 1.31)

II. Results				
	Smoking Category	Nominal	Median	90% U.I.
Risk per WLM (10^{-4})	ES	9.7	9.8	(4, 20)
	ES and NS	5.4	5.4	(2, 12)
Etiologic fraction	ES	0.12	0.11	(0.05, 0.3)
	ES and NS	0.136	0.12	(0.05, 0.3)
Years of life lost per radon-induced death	ES and NS	17.2	17.3	(15, 20)
Number of fatal lung cancer deaths from radon exposure	ES and NS	21,100	19000	(8,000, 45,000)
Exposure (WLM/y)	All	0.18	0.16	(0.12, 0.21)

Table 20: Monte Carlo simulation of *EF*, *YLL*, average residential exposure, and number of radon-induced fatal cancers with exposure factors fixed at nominal values.[a,b]

II. Results				
	Smoking Category	Nominal	Median	90% U.I.
Etiologic fraction	ES	0.12	0.12	(0.05,0.3)
	ES and NS	0.136	0.14	(0.06, 0.3)
Number of fatal lung cancer deaths from radon exposure	ES and NS	21,100	21,000	(9,000, 50,000)

[a] Radon concentration = 1.25 pCi/L, occupancy factor = 0.7, and equilibrium factor = 0.4.

[b] Same non-exposure parameter assumptions as in Table 19.

F. Uncertainty in Extrapolating to Low Exposure Rates

The BEIR VI Committee found that the ERR/WLM increased with decreasing exposure rate over the range of observation in the miner cohorts. The lowest exposure rate classification considered by the BEIR VI committee was <0.5 WL, which in the occupational context corresponds to annual exposures below 6 WLM. However, the average residential exposure rate is estimated to be only 0.181 WLM y^{-1}. Thus, applying the miner derived models to residential radon exposures necessitates an extrapolation to exposure rates well below the levels where there is useful epidemiological data upon which to base those models. This creates a source of uncertainty that is difficult to quantify, and we limit ourselves here to a qualitative discussion of the issue.

The increasing risk with decreasing exposure rate [inverse dose rate effect (IDRE)] observed in the miners parallels evidence from radiobiology, indicating that for a given dose of high-LET radiation, the effect is maximal at low dose rates. Typically, it is found that for sufficiently low doses the response is independent of dose rate, but that at high doses the response increases with decreasing dose rate. This characteristic behavior has been observed for cell transformation, produced by neutron or alpha-particle irradiation (Hill *et al.* 1982, Bettega *et al.* 1992). Moreover, such a dependence on dose rate has been observed in studies of lung cancer induction by radon decay products in rats (Chemaud *et al.* 1981, Cross *et al.* 1984).

The radiobiological evidence thus suggests that the ERR/WLM would be at least as high at the low exposure rate conditions prevailing in homes as the ERR/WLM derived from the miner studies. Indeed, there is no definitive evidence in BEIR VI that a low dose rate plateau had been reached in the lowest exposure rate category (cf. Table 3-3 in BEIR VI), so it could be argued that the risk in homes might be substantially underestimated by the BEIR VI model, which implicitly assumes that the ERR/WLM has already reached its maximum value at about 0.5 WL.

Biophysical explanations for the dose rate pattern described above for high-LET radiation generally involve saturation of damage to a radiosensitive population of cells. Because of this saturation phenomena, the effect of n hits to the same sensitive cell is less than n times the effect of 1 hit. The response could be increased, however, if the dose is protracted over a time scale comparable to the replenishment time of the sensitive cell population. A modified version of this mechanism has been proposed by Brenner and Sachs (2002) in which a small population of hypersensitive cells can be mutated by hits to neighboring cells (bystander effect). At higher doses the bystander effect becomes saturated and the process is dominated by direct hits to non-sensitive cells. It is also postulated that a direct hit to a sensitive cell usually kills that cell. The competition among these processes gives rise to a complex dose response relationship, in which the response rises rapidly to a maximum, then decreases, before beginning a further linear increase with dose. An IDRE would be present at intermediate dose levels.

Alternative biologically based models fit to the epidemiological data can yield very different predictions regarding the extrapolation to low exposure rates. Moolgavkar and colleagues have proposed a 2-stage model of carcinogenesis in which cells first undergo a single mutation that puts them in a precancerous, or intermediate state (Moolgavkar and Knudson 1981, Moolgavkar and Luebeck 1990). The pool of intermediate cells may then expand under the influence of cancer "promoters". Finally, a second mutational event can occur in an intermediate cell, which divides uncontrollably to form a malignant tumor. Applying this modeling approach to the analysis of the combined effects of cigarette smoking and radon on lung cancer incidence in the Colorado Plateau miners, Luebeck *et al.* (1999) concluded that radon acted mainly as a promoter of lung cancer and that it was its promoting activity which produced the observed IDRE. Further calculations with the two-stage model indicated that the ERR/WLM would not plateau with decreasing exposure rate, as expected from the findings discussed previously, but would peak and then fall off. It was also projected that the risks from residential radon exposure would be about 2 or 4 times lower than projected by the BEIR VI model, for smokers and never-smokers, respectively.

Two issues might be raised with respect to the conclusions derived from the two-stage model. First, there is the problem of basing the analysis solely on the Colorado Plateau miner data, with all its uncertainties in exposure estimation and the incompleteness of smoking information required for the analysis. Second, it attributes the IDRE to a promotional mechanism when there is only sketchy evidence that alpha-particle radiation acts as a cancer promoter, but there is ample evidence that alpha radiation is a mutagen and that the mutagenic effect exhibits an inverse dose rate dependence.

Bogen (1997) has proposed a variant of the two-stage model, which projects a protective effect of radon over a range of exposure rates, as suggested by the ecological studies of Cohen discussed earlier. At this point, all such models must be regarded as highly speculative. Only a more complete mechanistic understanding of alpha-particle induced carcinogenesis or more definitive epidemiologic data on the variation of lung cancer incidence with radon levels in homes can resolve the issue of exposure rate extrapolation.

G. Sensitivity Analysis of Risk Estimates to Assumptions about Health Effects from Exposures to Radon

The Monte Carlo simulation quantified uncertainties related to exposure factors and many of the parameters that were used for modeling excess relative risks. This section investigates the sensitivity of our risk estimates to assumptions about some factors that had not been accounted for in the Monte Carlo simulation. Examined here are: first, the sensitivity of our risk estimates to parameters that would differentiate risks for subgroups such as ES, NS, and children; and second, the dependency of the estimates on assumptions about the relationship between relative risks and time-since-exposure.

Let us first assume that ERRs are accurately represented by the submultiplicative scaled BEIR VI concentration model. Using β_{NS} and β_{ES} to denote risk coefficients for NS and ES, we have:

$$ERR\ (a)= \beta_{NS}\ (w_{5-14} + \theta_{15-24}\ w_{15-24} + \theta_{25+}\ w_{25+})\ \phi(a) \qquad \text{for NS}$$
$$ERR\ (a)= \beta_{ES}\ (w_{5-14} + \theta_{15-24}\ w_{15-24} + \theta_{25+}\ w_{25+})\ \phi(a) \qquad \text{for ES}$$

where β_{ES} = 0.9β, and β_{NS} = 2β, and β = 0.0634 (see equation 5 in Section VI.B.2).

Since most of the miners were ES, it is likely that the ratio (β_{ES}/β) is very close to 0.9. In contrast, there was much less data on NS, and as discussed in Section VII.B.3, there may be an extra factor of two uncertainty in the *ERR* for NS compared to that for the general population. Table 21 shows the risk per WLM and *EF* for β_{NS} = 0.0634 and β_{NS} = 0.254, corresponding to 0.5 and 2 times the nominal value for the scaled concentration model (β_{NS} = 0.127). For these calculations, time-since-exposure and attained age parameters were set to nominal values. Estimates of risk per WLM and *EF* for NS are proportional to the NS risk coefficient. For example, doubling the risk coefficient for NS (from 2β to 4β) doubles the (NS) estimates of risk per WLM (from 1.7×10^{-4} to 3.3×10^{-4}) and *EF* from (0.26 to 0.53). The effect on risk estimates for the entire population would naturally be much smaller: for NS risk coefficients of 2β to 4β, the risk per WLM would range from 5.4×10^{-4} to 6.3×10^{-4} and the *EF* would range from 0.14 to 0.16. Setting the NS risk coefficient to β would result in about an 8% reduction in the (overall) risk per WLM and *EF* estimates.

Table 21: Dependence of the risk per WLM and *EF* estimates on the NS risk coefficient

Estimate	Smoking Status	NS Risk Coefficient[a]		
		$\beta_{NS} = \beta$	$\beta_{NS} = 2\beta$	$\beta_{NS} = 4\beta$
Risk per WLM (10^{-4})	NS	0.8	1.7	3.3
	All	4.9	5.4	6.3
EF	NS	0.13	0.26	0.53
	All	0.12	0.14	0.16

[a] $\beta = 0.0634$ is the risk coefficient for the scaled concentration model. β_{NS} is the risk coefficient for NS.

Regarding the effect of childhood exposures, we defined β_c to be the exposure response parameter for exposures received before one's 18th birthday. Thus,

$$ERR\,(a) = \beta_c\,(w_{5\text{-}14,c} + \theta_{15\text{-}24}\,w_{15\text{-}24,c} + \theta_{25+}\,w_{25+,c})\,\phi(a)$$
$$+ \beta\,(w_{5\text{-}14,A} + \theta_{15\text{-}24}\,w_{15\text{-}24,A} + \theta_{25+}\,w_{25+,A})\,\phi(a)$$

where the subscript c denotes exposures received before the 18th birthday, and the subscript A denotes exposures received after the 18th birthday. The estimated risk per WLM from childhood exposures (exposures received before the 18th birthday) is proportional to β_c. For $\beta_c = \beta = 0.0634$, the estimated risk per WLM from childhood exposures is about 5.6×10^{-4}.

Table 22 shows the risk per WLM and *EF* (for lifetime exposures) for $\beta_c = 0.0317$ and $\beta_c = 0.127$, corresponding to 0.5 and 2.0 times the nominal value for the risk coefficient $\beta = 0.0634$. For these calculations, time-since-exposure and attained age parameters were again set to nominal values. Here, doubling the risk coefficient for children (from β to 2β) would increase the estimates of risk per WLM (from 5.4×10^{-4} to 6.7×10^{-4}) and *EF* (from 0.14 to 0.17) by about 24%. Setting the childhood risk coefficient to 0.5β would result in about a 12% reduction in the (overall) risk per WLM and a similar reduction in the *EF*.

Table 22: Dependence of the risk per WLM and *EF* estimates on the childhood risk coefficient (β_c)

Estimate	Childhood Risk Coefficient[a]		
	$\beta_c = 0.5\beta$	$\beta_c = \beta$	$\beta_c = 2\beta$
Risk per WLM (10^{-4})	4.7	5.4	6.7
EF	0.12	0.14	0.17

[a] $\beta = 0.0634$ is the risk coefficient for the scaled concentration model. The childhood risk coefficient, β_c, is the risk coefficient for exposures before the 18th birthday.

Finally, consider the sensitivity of the risk estimates to assumptions about the dependence of relative risk on time-since-exposure. For the scaled-concentration model, the relative risk (for a given attained age) plateaus – at 51% of the maximum value – after 25 years from time of exposure ($\theta_{25+} = .51$). However, the risk model can be generalized to incorporate the possibility that these relative risks continue to decline for time-since-exposures greater than 25 y. Suppose

$$ERR\,(a) = \beta\,(w_{5\text{-}14} + \theta_{15\text{-}24}\,w_{15\text{-}24} + \theta_{25\text{-}34}\,w_{25\text{-}34} + \theta_{35+}\,w_{35+})\,\phi(a)$$

where $\theta_{25\text{-}34}$ and θ_{35+} are time-since-exposure parameters for intervals 25 through 34 years or 35 years and greater. This is equivalent to the formulation used for our scaled-concentration model if $\theta_{25\text{-}34} = \theta_{35+}$. As shown in Table 23, if θ_{35+} were reduced by 50% so that $\theta_{35+} = 0.5 \times \theta_{25\text{-}34} = 0.255$, estimated risks would be about 20% smaller. The results of this sensitivity analysis are well within the range of plausible risk values based on results from the Monte Carlo simulation. Thus, although the scaled-concentration model does not incorporate all plausible ways in which risks depend on time-since-exposure, this particular "model" uncertainty does not seem to dominate other uncertainties that were quantified.

Table 23: Dependence of the estimated risk per WLM and *EF* estimates on assumptions on how relative risks fall off with time-since-exposure.

Estimate	Smoking Status	Time-since-exposure coefficient[a]	
		$\theta_{35+} = 0.5 \times \theta_{25\text{-}34}$	$\theta_{35+} = \theta_{25\text{-}34}$
Risk per WLM (10^{-4})	NS	1.3	1.7
	ES	7.8	9.7
	All	4.3	5.4
EF	NS	0.21	0.26
	ES	0.10	0.12
	All	0.11	0.14

[a] $\theta_{25\text{-}34}$ (equals 0.51) and θ_{35+} are time-since-exposure coefficients for the intervals 25 through 34 y and 35y or greater.

APPENDIX A: AGE-SPECIFIC, EVER-SMOKING PREVALENCE ESTIMATES

The calculation of gender- and age-specific ES estimates for 1990 was accomplished in three steps. The first step was to extrapolate white male and female prevalence estimates, obtained from the National Institutes of Health (NIH), for calendar years 1987 and 1988 to calendar year 1990. These NIH estimates were derived from results from six NHIS surveys (DHHS 1997), and were calculated for each of 17 different birth cohorts that range from 1885-89 to 1965-69. The second step adjusted these estimates using data from OSH and 1990 census data to obtain prevalence estimates for the entire male and female populations (across all races). Finally, smoothing splines were used to estimate the prevalence at each age. Details follow.

The first two steps of the process are illustrated in Tables A1 and A2. The third and fourth columns of the tables give the ES prevalence for whites estimated from six NHIS surveys for 1987 and 1988. We extrapolated the ES prevalence for each age group to obtain estimates in calendar year y by assuming a constant rate of change as follows:

$$ p(y) = p(y_0 - 1) \cdot \left(\frac{p(y_0)}{p(y_0 - 1)} \right)^{(y - y_0)} $$

Here $p(y)$ denotes the prevalence for a birth cohort in calendar year y, and y_0 denotes the last year for which we have NIH smoking prevalence estimates specific to that age group. For our purposes, $y = 1990$.

For example, the extrapolated white male ES prevalence for ages 20.5 to 25.5 y in 1990, is:

$$ 0.3705 = 0.3611 \, (0.3611 / 0.3565)^2 $$

To see this, note that for this age group, the last year for which we have NIH smoking prevalence estimates is 1988. Then since $y_0 = 1988$, $p(y_0) = 0.3611$, $p(y_0 - 1) = 0.3565$, and $y = 1990$, the result follows from the equation.

Next, OSH and 1990 census data were used to obtain prevalence estimates for the entire male and female populations (across all races). From OSH, the ES prevalence in 1990 was 58.7% for males and 42.3% for females. The 1990 census data allowed us to combine the ES prevalence estimates in the fifth column of Tables A1 and A2. Weighted averages of the 17 prevalence estimates, equal to 58.77% for males and 46.04% for females, were obtained using weights equal to the proportion of males and females (from the 1990 U.S. census) in each of the 17 age groups. Prevalence estimates for each age group were then obtained using the following formula:

$$p_i = \left(\frac{0.587}{0.5877}\right) p_{i,whites} \text{ for males}$$

$$p_i = \left(\frac{0.423}{0.4604}\right) p_{i,whites} \text{ for females}$$

Here, $p_{i,whites}$ denotes the ES prevalence for white males or females in the ith age group, and p_i denotes the ES prevalence for the entire population (all races). For example, the ES prevalence for males of ages 30.5 to 35.5 y is:

$$0.5285 = \left(\frac{0.587}{0.5877}\right) \cdot 0.5291$$

Finally, smoothing splines were used to obtain the age-specific ES prevalences given in Table A3. We accomplished this using a Newton-Raphson iterative procedure (see Hastie *et al.* 1990) for fitting the logits of the adjusted prevalence estimates in Tables A1 and A2. This involved the application of the MATLAB spine toolbox procedure "spaps" (de Boor 1998) to the logits, where the logits were input as functions of the midpoints (in years) of the corresponding age intervals; these midpoints are equal to 23, 28, ..., 103. Other inputs for this procedure were initial weights equal to the proportion of males or females alive in each age group (from 1990 census data), and tolerances set to 0.001 for females and 0.0003 for males.

Table A1: Ever-smoking prevalence estimates for males by age group.

Cohort (Birth year)	Age (years) on July 1, 1990	Ever-smoking prevalence (%)			
		Whites			Adjusted[b]
		1987	1988	1990[a]	1990
1965-69	20.5-25.5	35.65	36.11	37.05	37.00
1960-64	25.5-30.5	45.41	45.53	45.77	45.72
1955-59	30.5-35.5	52.91	52.91	52.91	52.85
1950-54	35.5-40.5	59.17	59.17	59.17	59.10
1945-49	40.5-45.5	66.25	66.31	66.43	66.35
1940-44	45.5-50.5	71.46	71.20	70.68	70.60
1935-39	50.5-55.5	73.29	73.02	72.48	72.40
1930-34	55.5-60.5	74.24	73.78	72.87	72.78
1925-29	60.5-65-5	76.77	76.39	75.64	75.55
1920-24	65.5-70.5	76.18	75.60	74.45	74.36
1915-19	70.5-75.5	74.72	74.13	72.96	72.88
1910-14	75.5-80.5	71.72	70.87	69.20	69.62
1905-09	80.5-85.5	66.80	65.75	63.70	63.62
1900-04	85.5-90.5	59.67	NA[c]	56.38	56.31
1895-99	90.5-95.5	NA	NA	50.56	50.50
1890-94	95.5-100.5	NA	NA	38.97	38.92
1885-89	100.5-105.5	NA	NA	39.09	39.04

[a] Extrapolated from 1987 and 1988 data, as discussed in text.
[b] Adjusted so that weighted average of age-grouped prevalence estimates equals OSH prevalence estimate of 58.7%.
[c] Prevalence estimates for cohorts born before 1900 were extrapolated using regression on logarithmically transformed prevalence data.

Table A2: Ever-smoking prevalence estimates for females by age group.

Cohort (Birth year)	Age (years) on July 1, 1990	Ever-smoking prevalence (%)			
		Whites			Adjusted[b]
		1987	1988	1990[a]	1990
1965-69	20.5-25.5	37.78	38.23	39.15	35.97
1960-64	25.5-30.5	46.51	46.60	46.78	42.98
1955-59	30.5-35.5	50.37	50.41	50.49	46.39
1950-54	35.5-40.5	47.87	48.04	48.38	44.45
1945-49	40.5-45.5	51.78	51.83	51.93	47.71
1940-44	45.5-50.5	55.77	55.60	55.26	50.77
1935-39	50.5-55.5	54.39	54.19	53.79	49.42
1930-34	55.5-60.5	53.00	52.78	52.34	48.09
1925-29	60.5-65-5	50.24	49.93	49.32	45.31
1920-24	65.5-70.5	46.34	45.92	45.09	41.43
1915-19	70.5-75.5	42.69	42.08	40.89	37.56
1910-14	75.5-80.5	34.86	33.96	32.23	29.61
1905-09	80.5-85.5	26.31	25.55	24.10	22.14
1900-04	85.5-90.5	16.64	NA[c]	14.98	13.76
1895-99	90.5-95.5	NA	NA	10.62	9.75
1890-94	95.5-100.5	NA	NA	7.00	6.43
1885-89	100.5-105.5	NA	NA	6.70	6.15

[a] Extrapolated from 1987 and 1988 data, as discussed in text.
[b] Adjusted so that weighted average of age-grouped prevalence estimates equals OSH prevalence estimate of 42.3%.
[c] Prevalence estimates for cohorts born before 1900 were extrapolated using regression on logarithmically transformed prevalence data.

Table A3: Smoothed age-specific ES prevalence estimates for males and females.

Age	Males	Females	Age	Males	Females	Age	Males	Females
18	0.2966	0.3075	48	0.7064	0.4974	77	0.7021	0.3118
19	0.3105	0.3169	49	0.7120	0.4991	78	0.6934	0.2966
20	0.3252	0.3276	50	0.7167	0.4997	79	0.6841	0.2813
21	0.3406	0.3392	51	0.7208	0.4994	80	0.6741	0.2658
22	0.3567	0.3516	52	0.7242	0.4983	81	0.6634	0.2504
23	0.3732	0.3644	53	0.7272	0.4966	82	0.6523	0.2351
24	0.3900	0.3774	54	0.7299	0.4943	83	0.6406	0.2201
25	0.4070	0.3902	55	0.7324	0.4915	84	0.6285	0.2055
26	0.4239	0.4025	56	0.7348	0.4882	85	0.6159	0.1914
27	0.4405	0.4141	57	0.7372	0.4844	86	0.6030	0.1778
28	0.4568	0.4245	58	0.7398	0.4803	87	0.5898	0.1649
29	0.4725	0.4334	59	0.7425	0.4757	88	0.5763	0.1526
30	0.4876	0.4409	60	0.7453	0.4707	89	0.5626	0.1411
31	0.5023	0.4469	61	0.7478	0.4653	90	0.5487	0.1303
32	0.5165	0.4513	62	0.7500	0.4595	91	0.5346	0.1202
33	0.5304	0.4542	63	0.7515	0.4533	92	0.5205	0.1107
34	0.5441	0.4556	64	0.7524	0.4468	93	0.5062	0.1019
35	0.5575	0.4561	65	0.7525	0.4399	94	0.4919	0.0938
36	0.5709	0.4561	66	0.7518	0.4327	95	0.4776	0.0862
37	0.5844	0.4563	67	0.7506	0.4250	96	0.4632	0.0792
38	0.5979	0.4573	68	0.7487	0.4168	97	0.4490	0.0727
39	0.6117	0.4595	69	0.7462	0.4082	98	0.4348	0.0667
40	0.6253	0.4627	70	0.7431	0.3990	99	0.4207	0.0611
41	0.6386	0.4668	71	0.7394	0.3891	100	0.4067	0.0560
42	0.6513	0.4714	72	0.7350	0.3783	101	0.3929	0.0513
43	0.6632	0.4764	73	0.7298	0.3667	102	0.3793	0.0470
44	0.6739	0.4815	74	0.7240	0.3541	103	0.3659	0.0430
45	0.6836	0.4864	75	0.7174	0.3406	104	0.3527	0.0394
46	0.6922	0.4909	76	0.7101	0.3265	105	0.3397	0.0360
47	0.6998	0.4946						

APPENDIX B: SMOOTHING THE BEIR VI RELATIVE RISK FUNCTIONS

As described in Part III, the BEIR VI preferred models specify that the excess relative risk (relative risk -1) depends on time-since-exposure, attained-age, and either rate of exposure (concentration) or duration according to the formula:

$$ERR = \beta \, (w_{5-14} + \theta_{15-24} \, w_{15-24} + \theta_{25+} \, w_{25+}) \phi_{age} \, \gamma_{z,}$$

The θ-parameters detail how relative risk depends on time-since-exposure, and ϕ_{age} describes the dependency on attained age. For almost all residential exposures, γ_z is equal to 1 in the concentration model; γ_z is equal to 13.6 in the duration model for attained ages greater than 40 y. β is constant for either model, and the effective exposure, $w_{eff} = w_{5-14} + \theta_{15-24} \, w_{15-24} + \theta_{25+} \, w_{25+}$. The effective exposure is a continuous function of attained age. However in the BEIR VI models, *ERR* is discontinuous at attained ages 55 y, 65 y, and 75 y because of discontinuities in the attained age function ϕ. For attained age categories <55 y, 55-65 y, 65-75 y, and > 75 y, corresponding values for ϕ are 1, 0.57, 0.34, and 0.28 for the duration model and 1, 0.65, 0.38, and 0.22 for the concentration model.

We smoothed the modeled *ERR* by using splines (Fritsch and Carlson, 1980) to smooth the attained age component ϕ. We did this by finding a monotonic spline with nodes at ages 40 y, 50 y, 55 y, 65 y, 75 y, 80 y, and 90 y for which the integral of ϕ was preserved for intervals 50-80 y, 55-65 y, and 65-75y. Results are given in Table B1.

Table B1: Spline smoothed values for ϕ^a, from the BEIR VI concentration and duration models.

Age (y)	Duration model	Concentration model	Age (y)	Duration model	Concentration model
50	1.0000	1.0000	66	0.2831	0.3277
51	0.9882	0.9888	67	0.2819	0.3154
52	0.9574	0.9599	68	0.2811	0.3055
53	0.9147	0.9197	69	0.2805	0.2971
54	0.8672	0.8752	70	0.2801	0.2896
55	0.8219	0.8329	71	0.2796	0.2822
56	0.7731	0.7892	72	0.2790	0.2742
57	0.7130	0.7377	73	0.2781	0.2647
58	0.6455	0.6808	74	0.2768	0.2530
59	0.5743	0.6211	75	0.2749	0.2383
60	0.5033	0.5610	76	0.2585	0.2124
61	0.4364	0.5031	77	0.2224	0.1743
62	0.3772	0.4498	78	0.1800	0.1344
63	0.3297	0.4038	79	0.1447	0.1028
64	0.2976	0.3674	80	0.1300	0.0900
65	0.2847	0.3432			

[a] This parameter describes the dependency of the modeled excess relative risks on attained age.

APPENDIX C: NOTATION AND FORMULAS

1) Summary risk measures

$RWLM$ = risk per WLM (working level month)
EF = etiologic fraction
YLL = average years of life lost per radon-induced lung cancer death

2) Basic quantities and functions

$h(x)$ = baseline lung cancer death rate at age x – the probability that a person, exposed to baseline levels of radon and alive at age x, will die from lung cancer before attaining age $(x + dx)$ is equal to $h(x)dx$

$S(x)$ = baseline survival function – the fraction of live-born individuals in a population exposed to baseline radon levels that is expected to survive to age x.

$g(x)$ = radon exposure rate (WLM / y) at age x.

$g_b(x)$ = background radon exposure rate set to 0.181 WLM / y.

$g_e(x)$ = radon exposure rate in excess of background
= $g(x) - g_b(x)$

r = lifetime risk of a premature lung cancer death due to excess radon exposure

$w(x)$ = cumulative radon exposure (WLM) at age x.

$e(x, g_e)$ = excess relative risk (ERR) at age x due to an excess radon exposure – the probability that a person, exposed to excess levels of radon and alive at age x, will die from lung cancer before attaining age $(x + dx)$ is equal to $h(x) [1 + e(x, g_e)] dx$.

$S(x, g_e)$ = fraction of live-born individuals in a population that are expected to survive to age x with excess radon exposure = $g_e(a)$ at ages $a \le x$

1.05 = the presumed sex ratio at birth (male-to-female)

3) Subscripts

Subscripts for $h(x)$, $S(x)$, $e(x)$, and $S(x, g)$ are used to denote specific populations or subpopulations:

pop nonstationary U.S. population that includes males, females, ES, and NS.

sta stationary population with fixed percentages of male ES, male NS, female ES and female NS

m subpopulation of males;   *f* subpopulation of females

ES subpopulation of ES;   *NS* subpopulation of NS

m, ES subpopulation of male ES, etc.

4) Baseline lung cancer death rates used in the text

$h(x)$ for a (sometimes generic) population at age x.
$h_{pop}(x)$ for the U.S. population that includes males, females, ES, and NS.
$h_{sta}(x)$ for a stationary population that includes males, females, ES, and NS.
$h_m(x)$ for males; $h_f(x)$ for females
$h_{ES}(x)$ for ES; $h_{NS}(x)$ for NS
$h_{m,ES}(x)$ for male ES, etc.

5) Baseline survival functions used in the text

$S(x)$, $S_{sta}(x)$, $S_m(x)$, $S_f(x)$, $S_{ES}(x)$, $S_{NS}(x)$, $S_{m,ES}(x)$, etc.

6) Survival functions modified by an excess radon exposure rate (g_e)

$S(x, g_e)$, $S_{sta}(x, g_e)$, $S_m(x, g_e)$, $S_f(x, g_e)$, $S_{ES}(x, g_e)$, $S_{NS}(x, g_e)$, $S_{m,ES}(x, g_e)$, etc.

We often assume that the excess radon exposure rate is equal to a very small constant, Δ, so that $g_e = \Delta$. The modified survival function, omitting subscripts, would be $S(x, \Delta)$.

7) Smoking prevalence

The fraction of live-born individuals that is expected to smoke at least 100 cigarettes during their lifetime is denoted by:

p_{birth} for the U.S. population (both males and females).
$p_{birth, m}$ for males; $p_{birth, f}$ for females

The fraction of individuals at age x that have smoked at least 100 cigarettes is denoted by:

$p(x)$ for the population
$p_m(x)$ for males; $p_f(x)$ for females

8) Excess relative risk (ERR)

$e_{ES}(x, g_e)$ for ES
$e_{NS}(x, g_e)$ for NS
$e_{ES}(x, g_e) = (0.9/2.0) \, e_{NS}(x, g_e)$

9) Survival function formulas

$$S(x) = (1.05\, S_m(x) + S_f(x)) / 2.05$$
$$S_m(x) = p_{birth,\, m}\; S_{m,ES}(x) + (1 - p_{birth,\, m})\; S_{m,NS}(x)$$
$$S_{ES}(x) = [1.05\, p_{birth,\, m}\; S_{m,ES}(x) + p_{birth,\, f}\; S_{f,\, ES}(x)] / p_{birth}$$

10) Adjustments to survival functions for smoking status

Notation:

$$A_{ES}(x) = S_{ES}(x) / S(x); \qquad A_{NS}(x) = S_{NS}(x) / S(x)$$
$$A_{m,ES}(x) = S_{m,\, ES}(x) / S_m(x); \quad A_{f,ES}(x) = S_{f,\, ES}(x) / S_f(x)$$
$$A_{m,NS}(x) = S_{m,\, NS}(x) / S_m(x); \quad A_{f,NS}(x) = S_{f,\, NS}(x) / S_f(x)$$

Equations:

$$A_{m,ES}(x) = exp\left(\int_0^x (h_m(a) - h_{m,ES}(a)) \cdot da \right)$$

11) Adjustments to survival functions due to radon exposure in excess of background

Notation:
$$B(x, g) = S(x, g) / S(x)$$
$$B_{m,ES}(x, g_e) = S_{m,ES}(x, g_e) / S_{m,ES}(x)$$

Formulas:

$$B_{m,ES}(x, g_e) = exp\left(\int_0^x (-e_{ES}(a, g_e) \cdot h_{m,ES}(a)) \cdot da \right)$$

12) Exposure to radon

$$g_b(x) = (1.25\ pCi/L)\, [(0.4)\, (10^{-2}\ WL(pCi/L)^{-1})]\, [(0.7)((365.25)(24)/170\ WLM/(WL\text{-}y))]$$

13) Life expectancy

Notation:

L = Life expectancy (note: this is similar to the notation used in BEIR IV).

L_m, L_f for males and females

L_{ES}, L_{NS} for ES and NS

$L_{m,ES}$ for male ES.

Formulas:

$$L_m = p_{birth,m} \cdot L_{m,ES} + (1 - p_{birth,m}) \cdot L_{m,NS}$$

$$L_f = p_{birth,f} \cdot L_{f,ES} + (1 - p_{birth,f}) \cdot L_{f,NS}$$

$$L_{ES} = 1.05 \cdot p_{birth,m} \cdot L_{m,ES} + p_{birth,f} \cdot L_{f,ES}$$

$$L_{NS} = 1.05 \cdot (1 - p_{birth,m}) \cdot L_{m,NS} + (1 - p_{birth,f}) \cdot L_{f,NS}$$

$$L = (1.05 \cdot L_m + L_f) / 2.05$$

14) Combining summary measures from different populations

Notation: Subscripts denote subpopulations based on gender and smoking.

$$RWLM_m = (p_{birth,m} \cdot L_{m,ES} \cdot RWLM_{m,ES} + (1 - p_{birth,m}) \cdot L_{m,ES} \cdot RWLM_{m,NS}) / L_m$$

$$RWLM_f = (p_{birth,f} \cdot L_{f,ES} \cdot RWLM_{f,ES} + (1 - p_{birth,f}) \cdot L_{f,NS} \cdot RWLM_{f,NS}) / L_f$$

$$RWLM_{ES} = (1.05 \cdot p_{birth,m} \cdot L_{m,ES} \cdot RWLM_{m,ES} + p_{birth,f} \cdot L_{f,ES} \cdot RWLM_{f,ES}) / (2.05 \, L_{ES})$$

$$RWLM_{NS} = \left[1.05(1 - p_{birth,m}) L_{m,NS} \cdot RWLM_{m,NS} + (1 - p_{birth,f}) L_{f,NS} \cdot RWLM_{f,NS} \right] / (2.05 \, L_{NS})$$

$$RWLM = (1.05 \cdot L_m \cdot RWLM_m + L_f \cdot RWLM_f) / (2.05 \, L)$$

$$EF = (1.05 \cdot R_{baseline,m} \cdot EF_m + R_{baseline,f} \cdot EF_f) / (2.05 \, R_{baseline})$$

APPENDIX D: LUNG CANCER RISKS BY RADON LEVEL AND SMOKING STATUS

Table D1 presents estimates of the risk of lung cancer death by radon level for NS, current smokers and the general population. Estimates are subject to considerable uncertainties as discussed in Sections VI.H, VI.I and Chapter VII. In particular, note that the risk models in the BEIR VI report did not specify excess relative risks for current smokers. Because the excess mortality rates are in some instances very large, baseline mortality rates were adjusted for the excess risk due to radon exposure.

Table D1: Lifetime risk of lung cancer death by radon level for never smokers, current smokers, and the general population.

Radon Level[a] (pCi/L)	Lifetime Risk of Lung Cancer Death from Radon Exposure in Homes		
	Never Smokers	Current Smokers	General Population
20	3.6%	26.3%	10.5%
10	1.8%	15.0%	5.6%
8	1.5%	12.0%	4.5%
4	0.7%	6.2%	2.3%
2	0.4%	3.2%	1.2%
1.25	0.2%	2.0%	0.7%
0.4	0.1%	0.6%	0.2%

[a] Assumes constant lifetime exposure in homes at these levels.
[b] Estimates are rounded to the nearest tenth of a percent. No indication of uncertainty should be inferred from this practice.

REFERENCES

Bergen, AW, N Caporaso. Cigarette smoking. *J. Natl. Cancer Inst.* **91**, 1365-1375, 1999.

Bettega, D, P Calzolari, GN. Chiorda and L Tallone-Lombardi. A transformation of C3H 10T½ cells with 4.3 MeV alpha particles at low doses: effects of single and fractionated doses. *Radiat. Res.* **131**, 66-71, 1992.

Bogen, KT. Do U.S. county data disprove linear no-threshold predictions of lung cancer risk for residential radon? — A preliminary assessment of biological plausibility. *Hum. Ecol. Risk Assess.* **3**, 157-186, 1997.

Brenner, DJ and RK Sachs. Do low dose-rate bystander effects influence radon risks? *Inter. J. Radiat. Biol.,* in press, 2002.

Cavallo, A. The radon equilibrium factor and comparative dosimetry in homes and mines. *Radiat. Prot. Dosim.* **92**, 295-298, 2000.

Cavallo, A. Reply to D. Krewski et al. Comment on protection of residential radon lung cancer risks: the BEIR VI risk models - (letter to the editor). *Radiat. Prot. Dosim.* **102**, 373, 2002.

(CDC) Centers for Disease Control, Morbidity and Mortality Weekly Report, Atlanta, Ga. - CDC Surveillance Summaries, 43 (SS-3). Surveillance for selected tobacco use behaviors - United States, 1900-94. Nov. 18, 1994. Also available at http://www.cdc.gov/nccdphp/osh/adstat1.htm

(CDC) Centers for Disease Control. Morbidity and Mortality Weekly Report, Atlanta, Ga. Cigarette smoking among adults—United States 1993. 1995.

(CDC) Centers for Disease Control. Morbidity and Mortality Weekly Report, Atlanta, Ga.- CDC Surveillance Summaries, 49 (SS-10) Youth tobacco surveillance – United States 1998-1999. Oct. 13, 2000.

Chemaud, J, R Perraud, R Masse and J Lafuma. Contribution of animal experimentation to the interpretation of human epidemiological data, pp. 551-557 in: *Proceedings of the International Conference on Radiation Hazards in Mining: Control, Measurement and Medical Aspects.* M. Gomez, ed., Kingsport Press Inc., Kingsport, TN, 1981.

Cohen, BL. A test of the linear no-threshold theory of radiation carcinogenesis. *Environmental Res.* **53**, 193-220, 1990.

Cohen, BL. Test of the linear no-threshold theory of radiation carcinogenesis for inhaled radon decay products. *Health Phys.* **68**, 157-174, 1995.

Cohen, BL. Response to Lubin's proposed explanations of our discrepancy. *Health Phys.* **75**, 18-22, 1998.

Cohen, BL. Response to criticisms of Smith et al. *Health Phys.* **75**, 23-28, 1998a.

Cross, FT. Invited commentary: residential radon risks from perspective of experimental animal studies. *Am. J. Epidemiol.* **140**, 333-339, 1994.

Cross, FT, RF Palmer, GE Dagle, RE Busch and RL Buschbom. Influence of radon daughter exposure rate, unattachment fraction, and disequilibrium on occurrence of lung tumors. *Radiat. Prot. Dosimetry* **7**, 381-384, 1984.

de Boor, C. *Spline Toolbox User's Guide.* The MathWorks, Inc., Natick, MA, 1998.

(DHHS) Department of Health and Human Services. A report of the Surgeon General: *Reducing the Health Consequences of Smoking. 25 Years of Progress.* U.S. Government Printing Office, Washington DC, 1989.

(DHHS) Department of Health and Human Services. *SEER Cancer Statistics Review 1973-1990.* National Institutes of Health 92-2789. 1995.

(DHHS) Department of Health and Human Services. Public Health Service and National Cancer Institute. Burns, DM , L Garfinkel, and JM Samet, editors. *Changes in Cigarette-Related Disease Risks and Their Implication for Prevention and Control.* NIH 97-4213. Smoking and Tobacco Control Monograph 8. U.S. GPO, Bethesda, MD, 1997.

Elkind, MM. Radon-induced cancer: a cell-based model of tumorigenesis due to protracted exposures. *Int. J. Radiat. Biol.* **66**, 649-653, 1994.

(EPA) Environmental Protection Agency. *Risk Assessments Methodology, Environmental Impact Statement: NESHAPS for Radionuclides. Background Information Document, Volume I.* EPA 520/1-89-005. U.S. Environmetal Protection Agency, Washington, DC, 1989.

(EPA) Environmental Protection Agency. *Technical Support Document for the 1992 Citizen's Guide to Radon.* EPA 400-R-92-011. U.S. Environmental Protection Agency, Washington, DC, 1992.

(EPA) Environmental Protection Agency. *Exposure Factors Handbook.* EPA/600/P-95/002/a-c. U.S. Environmental Protection Agency, Washington, DC, 1997.

Field, RW, BJ Smith and CF Lynch. Ecologic bias revisited, a rejoinder to Cohen's response to "Residential ^{222}Rn exposure and lung cancer: Testing the linear no-threshold theory with ecologic data." *Health Phys.* **75**, 31-33, 1998.

Fritsch, FN and RE Carlson. Monotonic piecewise cubic interpolation. *SIAM J. Numer. Anal.* **17**, 238-246, 1980.

George, AC and AJ Breslin. The distribution of ambient radon and radon daughters in residential buildings in the New Jersey–New York area, pp. 1272-1292 in: *Natural Radiation Environment III*, TF Gessell and WM Lowder, eds., U.S. DOE, 1980.

Goldsmith, JR. The residential radon-lung cancer association in U.S. Counties: a commentary. *Health Physics.* **76**, 553-557, 1999.

Greenland, S and J Robins. Review and commentary: Conceptual problems in the definition and interpretation of attributable fraction. *Am. J. Epidemiol.* **128**, 1185-97, 1988.

Greenland, S and J Robins. Invited commentary: Ecologic studies – biases, misconceptions and counterexamples. *Am. J. Epidemiol.* **139**, 747-760, 1994.

Hastie, TJ and RJ Tibshirani. *Generalized Additive Models*. Chapman and Hall, New York, 1990, p.96-98.

Hill, CK, FJ Buonoguro, CP Myers, A Han and MM Elkind. Fission-spectrum neutrons at reduced dose rate enhance neoplastic transformation. *Nature* **298**, 67-68, 1992.

Hopke, PK, B Jensen, CS Li, N Montassier, P Wasiolek, AJ Cavallo, K Gatsby, RH Socolow, AC James. Assessment of the exposure to and dose from radon decay products in normally occupied homes. *Environ. Sci. Technol.* **29**, 1359-1364, 1995.

Hornung, RW and TJ Meinhardt. Quantitative risk assessment of lung cancer in U.S. uranium miners. *Health Phys.* **52**, 417-430, 1987.

(ICRP) International Commission on Radiological Protection. *Protection Against Radon-222 at Home and at Work*. ICRP Publication 65. Pergamon, Tarrytown, NY, 1993.

Israeli, M. Deposition rates of Rn progeny in houses. *Health Phys.* **49**, 1069-1083, 1985.

James, AC. Dosimetry of radon and thoron exposures: Implications for risks from indoor exposures. Pp. 167-198 in: *Indoor Radon and Lung Cancer: Reality or Myth*, FT Cross ed., Battelle Press, Columbus, Ohio, 1992.

James, AC, A Birchall, and GH Akabani. Comparative Dosimetry of BEIR VI Revisited. *Radiation Protection Dosimetry*. Submitted, 2003.

Krewski, D, JH Lubin, JM Samet, PK Hopke, AC James, KP Brand. Projection of residential radon lung cancer risks: The BEIR VI risk models - (Letter to the editor) . *Radiat. Prot. Dosim.* **102**, 371-3, 2002.

Lubin, JH. On the discrepancy between epidemiologic studies in individuals of lung cancer and residential radon and Cohen's ecologic regression. *Health Phys.* **75**, 4-10, 1998.

Lubin, JH. Rejoinder: Cohen's response to "On the discrepancy between epidemiologic studies in individuals of lung cancer and residential radon and Cohen's ecologic regression." *Health Phys.* **75**, 29-30, 1998a.

Lubin, JH. The influence of residential radon exposure on the estimation of exposure-response trends for lung cancer in underground miners exposed to radon. *Radiat. Res.* **150**, 259-261, 1998.

Lubin, J. DISCUSSION: Indoor radon and risk of lung cancer. *Radiat. Res.* **151**, 105-106, 1999.

Lubin, JH and JD Boice, Jr. Lung cancer risk from residential radon: meta-analysis of eight epidemiologic studies. *J. Natl. Cancer Inst.* **89**, 49-57, 1997.

Lubin, JH, JD Boice Jr., C Edling, RW Hornung, G Howe, E Kunz, RA Kusiak, HI Morrison, EP Radford, JM Samet, M Tirmarche, A Woodward, SX Yao and DA Pierce. *Radon and Lung Cancer Risk: A Joint Analysis of 11 Underground Miners Studies.* National Institutes of Health, National Cancer Institute. NIH Publication No. 94-3644. U.S. Department of Health and Human Services, Washington, DC, 1994.

Luebeck, EG, WF Heidenreich, WD Hazelton, HG Paretzke and SH Moolgavkar. Biologically based analysis of the data for the Colorado uranium miners cohort: Age, dose and dose-rate effects. *Radiat. Res.* **152**, 339-351, 1999.

Malarcher, AM, J Schulman, LA Epstein, MJ Thun, P Mowery, B Pierce, L Escobedo, and GA Giovino. Methological Issues in Estimating Smoking-attributable Mortality in the United States. *Amer. J. Epidem.* **152**, 573-584, 2000.

Moolgavkar, SH and A Knudson. Mutation and cancer: a model for human carcinogenesis. *J. Natl. Cancer Inst.* **66**, 1037-1052, 1981.

Moolgavkar, SH and EG Luebeck. Two-event model for carcinogenesis: Biological, mathematical, and statistical considerations. *Risk Analysis* **10**, 323-341, 1990.

Marcinowski, F, RM Lucas and WM Yeager. National and regional distributions of airborne radon concentrations in U.S. homes. *Health Phys.* **66**, 699-706, 1994.

(NAS) National Academy of Sciences. *Health Risks of Radon and Other Internally Deposited Alpha-Emitters: BEIR IV.* National Academy Press, Washington, DC, 1988.

(NAS) National Academy of Sciences. *Comparative Dosimetry of Radon in Mines and Homes.* National Academy Press, Washington, DC, 1991.

(NAS) National Academy of Sciences. *Health Effects of Exposure to Radon: BEIR VI* National Academy Press, Washington, DC 1999.

(NCHS) National Center for Health Statistics. *Vital Statistics Mortality Data, Detail*, 1989. NTIS order number of datafile tapes: PB92-504554. U.S. Department of Health and Human Services, Public Health Service, Hyattsville, Md., 1992.

(NCHS) National Center for Health Statistics. *Vital Statistics Mortality Data, Detail*, 1990. NTIS order number of datafile tapes: PB93-504777. U.S. Department of Health and Human Services, Public Health Service, Hyattsville, Md., 1993a.

(NCHS) National Center for Health Statistics. *Vital Statistics Mortality Data, Detail, 1991.* NTIS order number of datafile tapes: PB93-506889. U.S. Department of Health and Human Services, Public Health Service, Hyattsville, Md., 1993b.

(NCHS) National Center for Health Statistics. *U.S. Decennial Life Tables for 1989-91. Vol. 1, No. 1. DHHS, PHS-98-1150-1*. United States Life Tables. Public Health Service, Washington, DC, 1997.

Nelson, CB, JS Puskin and DJ Pawel. Adjustments to the baseline lung cancer mortality for radon-induced lung cancers in the BEIR VI risk models. *Radiation Research*, **156**, 220-221, 2001.

(NJDEP) New Jersey Department of Environmental Protection. *Highlights of the Statewide Scientific Study of Radon,* New Jersey Department of Environmental Protection, Trenton, 1989.

Porstendörfer, J and A Reineking. Radon: Characteristics in Air and Dose Conversion Factors. *Health Phys.* **76**, 300-305, 1999.

Puskin, JS. Mortality rates with average county radon levels. *Health Phys.* (in press), 2003.

Rogers, RG and Powell-Griner EP. Life expectancies of cigarette smokers and nonsmokers in the United States. *Social Science in Medicine* **32**, 1151-59, 1991.

(SAB) Science Advisory Board. *An SAB Advisory: Assessing Risks from Indoor Radon.* An advisory prepared by the Radiation Advisory Committee (RAC) on Proposed EPA Methodology for Assessing Risks from Indoor Radon (EPA-SAB-RAC-ADV-99-010), Washington, DC, 1999

(SAB) Science Advisory Board. *An SAB Report: Assessment of Risks in Homes.* Review of the Draft Assessment of Risks from Radon in Homes by the Radiation Advisory Committee (RAC) of the Science Advisory Board (EPA-SAB-EC-00-010), Washington, DC, 2000

Samet, JM, DR Pathak, MV Morgan, DB Coultas, DS James and WC Hunt. Silicosis and lung cancer risk in underground uranium miners. *Health Phys.* **61**, 745-752, 1994.

Smith, BJ, RW Field and CF Lynch. Residential ^{222}Rn exposure and lung cancer: Testing the linear no-threshold theory with ecologic data. *Health Phys.* **75**, 11-17, 1998.

Thompson, DE, K Mabuchi, E Ron, M Soda, M Tokunaga, S Ochikubo, S Sugimoto, T Ikeda, M Terasaki, S Izumi and DL Preston. Cancer incidence in atomic bomb survivors. Part II: Solid Tumors, 1958-1987. *Radiat. Res.* **137**, S17-S67, 1994.

Tsang, AM and NE Klepeis. *Results Tables from a Detailed Analysis of the National Human Activity Pattern Survey (NHAPS) Response.* Draft report prepared for the U.S. Environmental Protection Agency by Lockheed Martin, 1996.

U.S. Census. 1990 Census data: Database C905TF3C1 from:
http://venus.census.gov/cdrom/lookup/1040151632

Xuan, XZ, JH Lubin, JY Li and WJ Blot. A cohort study in southern China of workers exposed to radon and radon decay products. *Health Phys.* **64**, 120-131, 1993.

Zang, EA and EL Wynder. Differences in lung cancer risk between men and women: examination of the evidence. *J. Natl. Cancer Inst.* **88**, 183-192, 1996.